ENERGY SCIENCE, ENGINEERING AND TECHNOLOGY

NEW DEVELOPMENTS IN SUSTAINABLE PETROLEUM ENGINEERING

ENERGY SCIENCE, ENGINEERING AND TECHNOLOGY

Additional books in this series can be found on Nova's website under the Series tab.

Additional E-books in this series can be found on Nova's website under the E-books tab.

ENERGY SCIENCE, ENGINEERING AND TECHNOLOGY

NEW DEVELOPMENTS IN SUSTAINABLE PETROLEUM ENGINEERING

RAFIQ ISLAM
EDITOR

Nova Science Publishers, Inc.
New York

Copyright © 2012 by Nova Science Publishers, Inc.

All rights reserved. No part of this book may be reproduced, stored in a retrieval system or transmitted in any form or by any means: electronic, electrostatic, magnetic, tape, mechanical photocopying, recording or otherwise without the written permission of the Publisher.

For permission to use material from this book please contact us:
Telephone 631-231-7269; Fax 631-231-8175
Web Site: http://www.novapublishers.com

NOTICE TO THE READER

The Publisher has taken reasonable care in the preparation of this book, but makes no expressed or implied warranty of any kind and assumes no responsibility for any errors or omissions. No liability is assumed for incidental or consequential damages in connection with or arising out of information contained in this book. The Publisher shall not be liable for any special, consequential, or exemplary damages resulting, in whole or in part, from the readers' use of, or reliance upon, this material. Any parts of this book based on government reports are so indicated and copyright is claimed for those parts to the extent applicable to compilations of such works.

Independent verification should be sought for any data, advice or recommendations contained in this book. In addition, no responsibility is assumed by the publisher for any injury and/or damage to persons or property arising from any methods, products, instructions, ideas or otherwise contained in this publication.

This publication is designed to provide accurate and authoritative information with regard to the subject matter covered herein. It is sold with the clear understanding that the Publisher is not engaged in rendering legal or any other professional services. If legal or any other expert assistance is required, the services of a competent person should be sought. FROM A DECLARATION OF PARTICIPANTS JOINTLY ADOPTED BY A COMMITTEE OF THE AMERICAN BAR ASSOCIATION AND A COMMITTEE OF PUBLISHERS.

Additional color graphics may be available in the e-book version of this book.

Library of Congress Cataloging-in-Publication Data

New developments in sustainable petroleum engineering / editor, Rafiq Islam.
     p. cm.
  Includes bibliographical references and index.
  ISBN 978-1-61324-159-2 (hardcover)
 1. Petroleum engineering. 2. Oil spills--Prevention. 3. Sustainable engineering. I. Islam, Rafiqul, 1959-
  TN870.N474 2011
  622'.3382--dc22
                              2011008472

Published by Nova Science Publishers, Inc. ✢ *New York*

CONTENTS

Preface vii

Chapter 1 Assessing Ecological Risks of Produced Water Discharge in Waves 1
Haibo Niu, Tahir Husain, Brian Veitch, Neil Bose, Kelly Hawboldt and Mukhtasor

Chapter 2 A Comprehensive Material Balance Equation with the Inclusion of Memory during Rock-Fluid Deformation 13
M. E. Hossain and M. R. Islam

Chapter 3 A Relation-Analysis-Based Approach for Assessing Risks of Petroleum-Contaminated Sites in Western Canada 35
Xiaosheng Qin, Bing Chen, Guohe Huang and Baiyu Zhang

Chapter 4 A Two-Step Modeling Approach for the Recovery of Free Phase LNAPL in Petroleum-Contaminated Groundwater Systems 53
Z. Chen and J. Yuan

Chapter 5 Improved Model for Predicting Formation Damage Induced by Oilfield Scales 65
Fadairo A. S. Adesina and Omole Olusegun

Chapter 6 Review of the Optimization Techniques in Groundwater Monitoring Network Design for Petroleum Contaminant Detection 79
Abdolnabi Abdeh-Kolahchi, Mysore G. Satish, Chefi Ketata and M. Rafiqul Islam

Chapter 7 Effect of Oilfield Sulphate Scales on the Productivity Index 99
Fadairo A. S. Adesina, C. T. Ako, Omole Olusegun and Falode Olugbenga

Chapter 8 A Monte-Carlo Simulation Based Assessment Approach for Analyzing Environmental Health Risks from B.T.E.-Contaminated Groundwater 109
L. Liu, J. B. Li and S. Y. Cheng

Chapter 9	Exact Solutions of Systems of Linear Integro-Differential Equations Using the HPM *Hossein Aminikhah and Jafar Biazar*	131
Chapter 10	Natural Additive for EOR Scheme during Chemical Flooding and its Environment Friendly Sustainable Application *M. Safiur Rahman and M. Rafiqul Islam*	141
Chapter 11	Characterization and Separation of Oil-in-Water Emulsion *Ajay Mandal, Pradeep Kumar, Keka Ojha and Subodh K. Maiti*	169
Index		181

PREFACE

Petroleum engineering is an engineering discipline concerned with the activities related to the production of hydrocarbons, which can be either crude oil or natural gas. Subsurface activities are deemed to fall within the upstream sector of the oil and gas industry, which are the activities of finding and producing hydrocarbons. This new book presents current research in the study of sustainable petroleum engineering including topics such as optimization techniques in groundwater monitoring network design for petroleum contaminant detection; a relation-analysis-based approach for assessing risks of petroleum-contaminated sites and an improved model for predicting formation damage induced by oilfield scales.

Chapter 1 – Ocean waves can generally increase the initial dilution of produced water and subsequently affect the ecological risk. To study the wave effects on ecological risk, a probabilistic based buoyant jet dispersion model has been integrated with risk analysis software, @ RISK, to assess ecological risk for a case corresponding to published parameters for the White Rose field, off east coast of Canada. The ecological risks on sheepshead minnows were evaluated by conducting Monte Carlo simulation for four different discharge scenarios. The Hazard Quotients (HQs) method was used to characterize the risk. Contour plots of probability distributions of HQs from those four discharge configurations were compared and discussed. The wave effects were identified by comparison of the wave and no wave simulations.

Chapter 2 – With the advent of fast computational techniques, it is time to include all salient features of the material balance equation (MBE). The inclusion of time dependent porosity and permeability can enhance the quality of oil recovery predictions to a great extent. This alteration occurs due to change of pressure and temperature of the reservoir which causes a continuous change of rock-fluid properties with time. However, few studies report such alterations and their consequences. This study investigates the effects of permeability, pore volume, and porosity with time during the production of oil. Moreover, a comprehensive MBE is presented. The equation contains a stress-strain formulation that is applicable to both rock and fluid. In addition, this formulation includes memory effect of fluids in terms of a continuous time function. Similar time dependence is also invoked to the rock strain-stress relationship. This formulation results in a highly nonlinear MBE, with a number of coefficients that are inherently nonlinear (mainly due to continuous dependence on time). This enhances the applicability of the model to fractured formations with dynamic features. Such formulation is different from previous approach that added a transient term to a steady state equation. For selected cases, the MBE is solved numerically with a newly developed nonlinear solver. A comparative study is presented using field data. Results are compared with conventional MBE approach. This comparison highlights the improvement in the recovery factor (RF) as much as 5%. This new version of MBE is applicable in the cases where rock/fluid compressibilities are available as a function of pressure from laboratory measurements or correlations. It is also applicable where non-pay zones are active as connate water and solid rock expansions are strong enough with pressure depletion. Finally, suitability of the new formulation is shown for a wide range of applications in petroleum reservoir engineering.

Chapter 3 – Effective reflection of and coping with uncertainties are essential for generating reliable risk assessment outcomes. In this study, an integrated risk assessment approach was proposed for assessing environmental risks associated with contamination of multi-component petroleum hydrocarbons. The approach consists of (a) predicting contaminant flow and transport through a multiphase multi-component numerical modelling system; (b) using interval-anlaysis approach to analyze effects of uncertainties associated with subsurface conditions; and (c) applying fuzzy relation analysis to quantify the general risks based on the interval-analysis results. The application of the proposed approach to a petroleum-contaminated site in western Canada indicated that system uncertainties would have significant impact on the risk assessment results. Implementation of risk assessment under more deterministic conditions generates clearer risk assessment outcomes, but leads to missing of more valuable information; conversely, risk assessment under more uncertain circumstances offers more comprehensive information, but leads to higher vagueness of risk descriptions. Application of the proposed approach in risk assessment of groundwater contamination represents a new contribution to the area of petroleum waste management under uncertainty. It is not only useful for evaluating risks of a system containing multiple factors with complicated interrelationships, but also advantageous in situations when probabilistic information is unavailable for performing a conventional stochastic risk assessment.

Chapter 4 – This study presents a two-step modeling method to the recovery of leaked petroleum product in groundwater system. Specifically, an oil volume estimation method is developed to calculate the total volume of LNAPL residing in both saturated and unsaturated zones under concern. With the information of LNAPL distribution in the groundwater system, porous media multiphase simulation technique is then used to examine the remediation alternative of vacuum-enhanced multiphase extraction (i.e., VER). VER modeling results can be also verified by the oil estimation analysis through comparing recovered and estimated LNAPL volumes. At the study site, the soil and groundwater are contaminated by leaked condensate from perforated underground storage tanks. The VER system designed for the separation and recovery of condensate from subsurface consists of one extraction well. Good results have been obtained with the recovered amount of LNAPL close to the estimated volume. It indicates the developed method is effective in simulating the recovery process of LNAPL from the contaminated site, and thus provides optimal parameters for the remediation program.

Chapter 5 – The process of formation damage due to oilfield scale precipitation and accumulation has been quantitatively modelled based on existing thermodynamics and deposition kinetic model. A variety of models on formation damage due to solid precipitation in porous media during water flooding have been reported in the literatures. Early models were based on chemical reaction involving dissolution/precipitation while neglecting the effect of operational and reservoir/brine parameters. This paper presents modified models for predicting permeability damage due to oilfield scale precipitation. The key operational and reservoir parameters which influence the magnitude of flow impairment by scale deposition were identified through the modification.

Chapter 6 – In order to detect petroleum contaminants in groundwater, it is paramount to design a proper monitoring network. Groundwater monitoring network design has advanced in recent years with the Genetic Algorithms optimisation techniques to address and achieve the optimal management strategy. This paper presents and describes various optimisation techniques, especially the state of the art robust Genetic Algorithms and their application in groundwater management focused on groundwater monitoring network. This review describes the Optimisation of groundwater management has become an active area of research in the last several years because of its ability to reduce cost substantially. It is the purpose of this paper to provide a comparison of a variety of optimisation methods on a groundwater problem due to petroleum contamination.

Chapter 7 – The precipitation and deposition of scale pose serious injectivity and productivity problems. Several models have been developed for predicting oilfield scales formation and their effect on deliverability of the reservoir to aid in planning appropriate injection water programme. In this study an analytical model has been developed for predicting productivity index of reservoir with incidence of scale deposition in the vicinity of the well bore.

Chapter 8 – In North America, numerous aquifers as fresh drinking water supply sources have been contaminated from various sources such as septic systems, leaking underground storage tanks, spills or improper disposal of industrial chemicals, and leachate from solid and hazardous waste landfills. A major task associated with the contamination of aquifers is to develop effective tools to assess and determine the health risks to individuals potentially consuming the water from these aquifers. This poses many obstacles due to the complexity of the environment that the contaminant is spilled in and the population consuming the water. This paper presents a Monte Carlo simulation based methodology which is capable of considering many variations present in the natural environment and human populations. The methodology is tested using a hypothetical aquifer and population case. Downgradient contaminant concentrations resulting from a leaking underground storage tank, containing petroleum, are calculated using an analytical solution to the two-dimensional advection-dispersion transport model. Contaminants under consideration include benzene, toluene, and ethylbenzene (BTE) which can cause deleterious health effects. Parameters and variables in the model are considered as random numbers. Contaminant ingestion dose is calculated using the stochastic exposure-dose model. Random population variables are used to give a distribution for contaminant ingestion dose for each contaminant. The chronic non-cancer hazard index is used to determine the risk associated with the ingestion of non-carcinogenic contaminants. The cancer risk is calculated using the slope factor, given by the USEPA, for the carcinogenic contaminants. Results of the case study indicate that environmental health risks can be effectively analyzed through the developed methodology. They are useful for supporting the related risk-management and remediation decisions.

Chapter 9 – In this paper, author introduce a new version of homotopy perturbation method to obtain exact solutions of systems of linear integro-differential equations. Theoretical considerations are discussed. Some examples are presented, to illustrating the efficiently and simplicity of the method.

Chapter 10 – Alkali and alkali/polymer solutions are well known techniques for the chemical flooding application. For this scheme, synthetic high-pH alkaline solutions are commonly used. These solutions are not environment friendly and are expensive. As a result, alkaline flooding has lost its appeal in last few decades. However, low-cost, environment-friendly alkaline solutions hold promises. This paper demonstrates how wood ash can be used as a source of low-cost alkali that is also environment friendly. The feasibility of using high pH alkaline solution, extracted from wood ash was conducted in the laboratory. From the experimental studies, it was found that the resulting solution was transparent and had high alkalinity. It was also found that the pH value of 6% wood ash-extracted solution was very close to the pH value of 0.5% synthetic sodium hydroxide or of 0.75% synthetic sodium meta silicate solution. A preliminary microscopic study of oil/oil droplets interaction in natural alkaline solution was carried out in order to understand the oil/water interface changes with time and its effect on oil/oil droplet coalescence. The microscopic study showed that two oil droplets were coalesced after 3.5 minute in 6% wood ash extracted solution. The interaction of the alkali in floodwater and the acids in reservoir crude oil result in the in-situ formation of surfactants that causes the lowering of interfacial tension (IFT) in caustic flooding that assist in the oil recovery process by mobilizing oil. In this study, the interfacial tension was measured using the Du Nouy ring method and it was observed that this environment friendly alkaline solution effectively reduced the interfacial tension with the acidic crude oil. Characterization of maple wood ashes has been was investigated using a variety of techniques, including, SEM-EDX, XRD, NMR. The SEM micrographs of the maple wood ash samples showed that the ash samples consisted of some porous and amorphous particles of carbon and several inorganic particles of irregular shape. The X-ray analysis on maple wood ash revealed that the predominant elements in the wood ash samples were oxygen, calcium, potassium, silicon. Lesser amounts of the elements were sodium, magnesium, titanium and aluminum also observed in maple wood ashes. The XRD analysis revealed that the major components of maple wood ashes were calcium oxide, potassium oxide, manganese oxide, silica oxide and magnesium oxide which were alkaline in nature. The ^{13}C CP/MAS NMR spectrum of maple wood ashes showed a very pronounced intense peak around 168.36 ppm was revealed that was assigned for the carbonate, [(O)$_2$-C=O]. It was revealed that nutrient elements status in fresh and treated maple wood ashes is almost same. Therefore, after alkaline extraction for EOR application, the same maple wood ashes has potential to use as a source of nutrient to soil and plants.

Chapter 11 – Chemical demulsification is most widely applied in petroleum industries, painting and wastewater treatment technology and involves the use of chemical additives to accelerate the emulsion breaking process. The stability of the emulsion has been characterized and it is observed that the stability depend on oil-water contact time, turbulence and amount of oil in contact with the water. A series of experiments have been carried out with different demulsifiers for separation of oil from oil-in-water emulsion. More than 90% separations are obtained with some demulsifiers under specific operating conditions.

Versions of these chapters were also published in *Advances in Sustainable Petroleum Engineering Science*, Volume 1, Numbers 1-4, edited by Rafiq Islam, published by Nova Science Publishers, Inc. They were submitted for appropriate modifications in an effort to encourage wider dissemination of research.

Chapter 1

ASSESSING ECOLOGICAL RISKS OF PRODUCED WATER DISCHARGE IN WAVES

Haibo Niu[1]*, Tahir Husain[1], Brian Veitch[1], Neil Bose[2], Kelly Hawboldt[1] and Mukhtasor[3]

[1]Faculty of Engineering and Applied Science,
Memorial University of Newfoundland, St. John's, NL, Canada
[2]Australian Maritime College, Launceston, Tasmania, Australia
[3]Faculty of Ocean Technology, Sepuluh Nopember Institute of Technology,
ITS, Surabaya, Indonesia

ABSTRACT

Ocean waves can generally increase the initial dilution of produced water and subsequently affect the ecological risk. To study the wave effects on ecological risk, a probabilistic based buoyant jet dispersion model has been integrated with risk analysis software, @ RISK, to assess ecological risk for a case corresponding to published parameters for the White Rose field, off east coast of Canada. The ecological risks on sheepshead minnows were evaluated by conducting Monte Carlo simulation for four different discharge scenarios. The Hazard Quotients (HQs) method was used to characterize the risk. Contour plots of probability distributions of HQs from those four discharge configurations were compared and discussed. The wave effects were identified by comparison of the wave and no wave simulations.

Keywords: produced water; hydrodynamic model; ecological risk; hazard quotients; waves.

1. INTRODUCTION

The recent increase in offshore oil and gas development off the east coast of Canada has created concern regarding the capacity of the marine environment as an intermediate buffer

---

* Corresponding author: e-mail: hniu@mun.ca

zone for receiving the produced water and the subsequent mixing of the waste materials with offshore waters. Produced water, which is normally comprised of formation water and injected seawater, is the largest waste stream associated with offshore petroleum production and its composition is strongly site dependent. The risks associated with the offshore discharge of produced waters depend on the composition of contaminants in the water and their distribution in the receiving environment. The distribution of produced waters in the receiving environment depends upon the discharge and ambient parameters, such as discharge depth, pipe orientation, relative density, current speeds, the presence of stratified density layers and the wave conditions. For discharge at given ambient conditions, the environmental concentration and therefore ecological risks may be reduced with optimally designed outfalls.

The effects of ocean outfall design on the ecological risks of produced water have been studied by Mukhtasor (2001), who introduced a framework of ecological risk-based design and conducted a case study based on the Terra Nova oil field, about 350 km off the east coast of Canada. The framework of ecological risk-based design requires the integration of hydrodynamic modeling and ecological risk assessment. The limitation of Mukhtasor's (2001) work is the weakness in the hydrodynamic model. The model can only simulate vertical discharges with positive buoyancy, which means the direction of the buoyancy must be the same as the discharge direction. In practice, many produced water discharges involve negative buoyancy, which means the direction of the buoyancy is different from the discharge direction. Therefore, in order to evaluate the ecological risks resulting from negatively buoyant discharges, the hydrodynamic model of Mukhtasor (2001) must be improved.

Previous studies of negatively buoyant jets (Zeitoun *et al.*, 1970; Robert and Toms, 1987; Roberts *et al.*, 1997) have proved that the discharge angle is a key parameter which has a significant impact on the initial dilution. As a result, the discharge angle may have significant effects on ecological risks.

It has been shown by previous studies (Chin, 1987; Hwung *et al.*, 1994; Chyan *et al.*, 2002) that surface waves can significantly increase the dilution achieved by ocean outfalls. Using data obtained from an experiment conducted off Sydney, Australia, Tate (2002) also concluded that high frequency internal waves can increase the initial dilution of buoyant jets by a factor exceeding two. Because of this effect, the wave is expected to reduce the ecological risk of produced water.

In this study, the effects of wave on the ecological risk of a negatively buoyant produced water discharge will be investigated by conducting a case study.

2. STUDY AREA

White Rose is an oil field located 350 km off the east coast of Newfoundland. The estimated maximum flowrate of produced water is 30,000 m^3/day (0.35 m^3/s), and the estimated density and temperature of the produced water are 25 ppt and 60 °C (Hodgins and Hodgins, 2000). Water depths in the study area range from about 120 m to 130 m. Currents in the vicinity of White Rose are dominated by wind and tide, with a mean flow to the south. The salinity of sea water is around 32.2 ppt and the water temperatures vary from 14.6 °C in summer to 0.25 °C in winter.

The density of the produced water is less than that of the ambient seawater. Once it is discharged, ecological entities, particularly those in the water column in the vicinity of the discharge, are potentially exposed to it. The potential risks include inhibition of growth and survival of fish (e.g. capelin, mackerel, cod and tuna) and shellfish (e.g. northern shrimp and snow crab) species.

Hodgins and Hodgins (2000) have calculated the exposure concentration for this site by assuming discharge of the water downward from a 0.356 m diameter pipe at 3 meter depth. In the following sections, ecological risks resulting from this discharge configuration will be calculated first. The effects of discharge configuration variations and wave effects on ecological risks will then be evaluated.

3. METHODOLOGY

3.1. Dispersion Model

To assess the risk, a hydrodynamic model is needed to calculate the exposure concentration. The model employed in this study has five sub-modules: (1) initial dilution model, (2) wave effect model, (3) boil location model, (4) intermediate model, and (5) far field mixing model.

The initial dilution model is an inclined (or declined) dense jet model as described by Roberts and Toms (1987) and Roberts *et al.* (1997). As shown in Figure. 1, a nozzle is declined downward to the horizontal at an angle of θ and discharge is through a round nozzle of diameter D at velocity U_j. The y_t is the maximum rise (or descent height), the y_L is the thickness of surface layer, S_i is the dilution at the boil point, x_i is distance from discharge point to boil point, x_m is the length of initial mixing zone, and S_m is the dilution at the end of initial mixing zone.

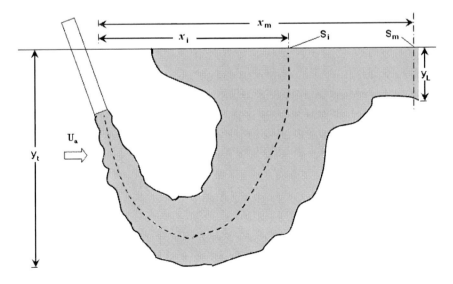

Figure 1. Definition Sketch for the Discharge.

The equations to determine above parameters are given by Robert *et al.* (1997) as

$$S_i = C_i F_i \tag{1}$$

$$S_m = C_m F_r \tag{2}$$

$$y_t = C_t D F_r \tag{3}$$

$$y_L = C_L D F_r \tag{4}$$

where the C_i, C_m, C_t, and C_L are experimentally determined constants. F_r is the densimetric Froude number.

Although the wave effect on ecological risk has not been has not been studied, the effects of waves on initial dilution has been investigated (Chin, 1987; Hwung *et al.*, 1994; Chyan *et al.*, 2002). The wave effects on initial dilution can be characterized by

$$\frac{S_{wave}}{S_{nowave}} = 1 + C_W \frac{L_q}{Z_m} + \varepsilon_2 \tag{5}$$

where S_{wave} is the initial dilution with the influence of waves, S_{nowave} is the initial dilution without the influence of waves, C_W is an experimentally determined coefficient that depends on the discharge angle (its value is given later), ε_2 is a random quantity normally distributed with a mean of zero and a standard deviation of 0.089, L_q is a length scale that measures the length over which the port geometry influences the effluent behavior, and Z_m is a length scale that measures the distance required for the jet momentum to be on the order of the wave induced momentum.

The far field dilution of a buoyant plume is governed by buoyant spreading and oceanic turbulent diffusion. Buoyant spreading is a horizontal transverse spreading and vertical collapse of the plume due to the residual buoyancy contained in the plume, while turbulent diffusion is a passive dispersion resulting from oceanic eddies or turbulence. Generally, both buoyant spreading and turbulent diffusion could be important at a distance from an outfall. As the buoyancy effects gradually diminish as the plume travels downstream, only the ambient turbulence is of concern after the transition point.

A buoyant spreading model (Huang *et al.*, 1996) has been adopted in this study. The details of the buoyant spreading model and the associated uncertainty measurements have been described by Mukhtasor (2001) and the turbulent diffusion model is given in Equation 6:

$$C(x,y) = \frac{1}{2} C_{BS} \left[erf\left(\frac{y + W_{BS}/2}{\sqrt{2}\sigma(x)} \right) - erf\left(\frac{y - W_{BS}/2}{\sqrt{2}\sigma(x)} \right) \right] \tag{6}$$

where $C(x,y)$ is the concentration at a point x, y downstream, C_{BS} is the centerline concentration at the end of buoyant spreading, W_{BS} is the plume width at the end of buoyant spreading, $\sigma^2(x)$ is the variance of the concentration distribution which can be related to the diffusion length scale L_y by $L_y = 2(3)^{1/2}\sigma(x)$. To estimate the concentration, the $\sigma(x)$ can be obtained through field dye diffusion experiments, or be estimated from empirical ocean diffusion diagrams in the case of lack of experimental data. Three diffusion laws (Fickian, Linear, and Inertial sub-range) are normally used to describe the turbulent processes. However, empirical results from large lakes and oceans show that natural turbulent processes cannot always be described by any one single diffusion law. Therefore, an empirical diffusion law (Equation 7), which was developed based on extensive oceanic diffusion studies (Okubo, 1971), is used in this work.

$$\sigma(x) = kT^f \qquad (7)$$

where T is the effective diffusion time approximated by x/U_a, k and f are regression coefficients that depend on the time scale (day, week, or month) or length scale of dispersion. For the acute effects of produced water in the vicinity of the platform, a time scale of one day is important because the produced water can travel several kilometres during one day and become well diluted. It is assumed that there is no vertical variation of concentration and only the horizontal diffusivity was considered in this work. The limitation of this assumption is that it will underestimate the dilution in the buoyant spreading layer and ignore the presence of contaminants at other depths of water column.

3.2. Probabilistic-Based Modeling and Ecological Risk Assessment

For computation of exposure concentration using dispersion models, two approaches can be used: worst-case approach and probabilistic based approach.

The worst case approach calculates a single value exposure concentration by considering the combination of parameters at worst case conditions. The advantage of the worst-case approach is its simplicity. However, the results derived from this approach may be more stringent than necessary because the approach is conservative.

Unlike the worst-case approach, which recognizes the uncertainties but does not model it explicitly, a probabilistic approach considers parameter variability, which is often described in terms of time series or probability distributions. This approach is often implemented using a Monte Carlo simulation method and the result is a probabilistic description of concentrations. Several applications of probabilistic based assessment of effluent discharges into rivers have been reported (Bumgardner et al., 1993; Donigian and Waggy, 1974). Huang et al. (1996) used this approach to model a sewage ocean outfall. More recently, this approach was adopted by Mukhtasor (2001) to model the dispersion of produced water in marine environment.

To calculate the probabilistic distribution of exposure concentration, the method of Mukhtasor (2001) was employed in this study. The probability distributions of parameters in equations (1) to (7) were first defined. Monte Carlo simulations were then performed to generate a random number for each parameter from its associated distribution. An exposure concentration can be obtained using the generated values. After this procedure is repeated a sufficient number of times (for example, 1000 times), the probabilistic distribution of exposure concentration can be calculated.

With the computed exposure concentration, ecological risks can then be characterized. Ecological risks may be described qualitatively or quantitatively. For a quantitative approach, basically, there are two methods available: quotient method and continuous exposure-response method. The quotient method has proved to be an effective method in the analysis of risks from produced water discharges (Mukhtasor, 2001) and was adopted in this study.

The quotient or hazard quotient (HQ) method is based on chronic benchmark concentrations of the whole effluent toxicity to individual species. The HQ can be expressed as:

$$\text{Quotient} = \frac{EC}{BC} \qquad (8)$$

where EC stands for the exposure concentration, and BC is the benchmark concentration.

Similar to the parameters in equations (1) to (7), the variability of BC in equation (8) can also be considered by defining a probability distribution (for example, lognormal) in the probabilistic analysis. The output of equation (8) is a probabilistic distribution of ecological risk.

3.3. Simulation of Ecological Risks

As described in the previous section, the exposure concentration for this study site has previously been studied by Hodgins and Hodgins (2000). The discharge of Hodgins and Hodgins (2000) is 90° from horizontal. In this study, a discharge angle of 60° rather than 90° was used to maximize the initial dilution as suggested by Roberts *et al.* (1997). To evaluate the effects of waves on ecological risk, the existing discharge scenario ($D = 0.356$ m) at the maximum flowrate (0.35 m^3/s) was evaluated first. To get more conclusive results on wave effects, discharges from different sizes of pipes at different flow rates were also evaluated. The estimated early stage discharge rate is 0.11m^3/s. Four different discharge scenarios were used by combining the maximum and early stage flow rates with two possible pipe sizes, 0.356m and 0.3m in diameter.

An influence area of 200 m × 200 m around the FPSO was studied. The coordinate system was defined using the same method as Huang *et al.* (1996). The Monte Carlo Simulation (MCS) method with a Latin Hypercube Sampling (LHS) was used in this work to consider the uncertainty associated with the variability of model inputs. The uncertainty measures associated with the models are listed in Table 1. The ambient water data for this area were analyzed and it was found that the ambient current speed can be fitted by a Gamma distribution and their directions can be approximated by a Beta distribution (see Table 1). For a given simulation, a sample is randomly drawn from each distribution and the value is assigned to the model input. The concentrations for the grid points within the study area were then calculated using the model described above. After repeating the simulation a given number of times, the concentration distribution for the whole study area was obtained. The simulation was conducted using risk analysis software, @RISK.

The No Observed Effect Concentrations (NOECs) data from Meinhold *et al.* (1996) for sheepshead minnows (Cyprinodon variegatus), a euryhaline fish, were used as the measurement endpoint. The reason for using sheepshead minnows is the availability of toxicity study data.

Table 1. Uncertainty measure of model inputs

Parameter	Value	Distribution
C_i	1.6±12%	Triangle
C_m	2.6±15%	Triangle
C_W	5.67±0.64	Normal
k	0.06±0.02	Normal
f	2.25±0.04	Normal
Ambient density (kg/m^3)	1001.63	Constant
Effluent density (kg/m^3)	1024.84~1025.81	Uniform
Current Speeds (m/s)	0.002~0.515	Gamma
Current Direction (Radian)	0.03~6.28	Beta
L_q/Z_m	0~0.15	Uniform

4. RESULTS AND DISCUSSIONS

An example of the distribution of mean and 95 percentile concentrations (1000 simulations) is plotted in Figure. 2. With this exposure concentration, the ecological risks were analyzed by determining the relationships between the exposure to the contaminant and the effects on the measurement endpoint (e.g. survival, growth), based on results of toxicity studies.

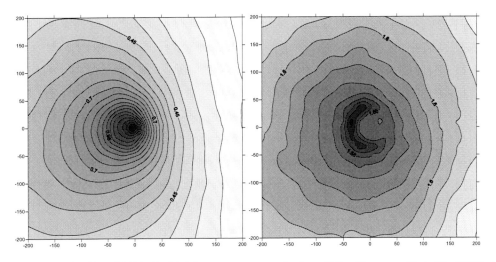

Figure 2. Mean (Left) and 95%-tile (right) exposure concentrations (%) with wave effects (D=0.356m).

By assuming a lognormal distribution of benchmark concentration, Monte Carlo. By assuming a lognormal distribution of benchmark concentration, Monte Carlo simulations were performed and the results of the HQs are shown in Figures 3 to 6.

It can be seen from Figures 3 to 6 that ocean surface waves have a significant effect on the predicted ecological risk of produced water. For the discharges at the maximum flow rate of 0.35 m^3/s through the 0.356 m pipe, if the waves are ignored, the impact area with an HQ greater than unity is approximately 7850 m^2 (radius of 100 m). However, if the wave effect is considered, the area for an HQ greater than unity is very small (hard to distinguish from Figure. 3). The same trend can be found from Figure. 4 for the discharges through a 0.3 m pipe.

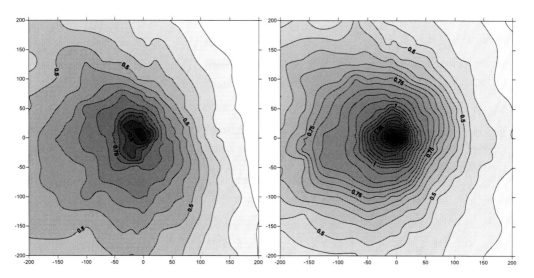

Figure 3. 95%-tile HQ for the survival of fish from simulations with (left) and without (right) wave effects (D= 0.356 m, Q = 0.35m^3/s).

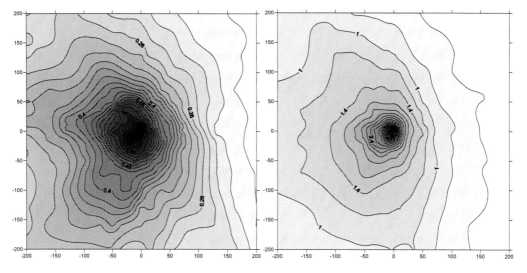

Figure 4. 95%-tile HQ for the survival of fish from simulations with (left) and without (right) wave effects (D=0.3 m, Q = 0.35m^3/s).

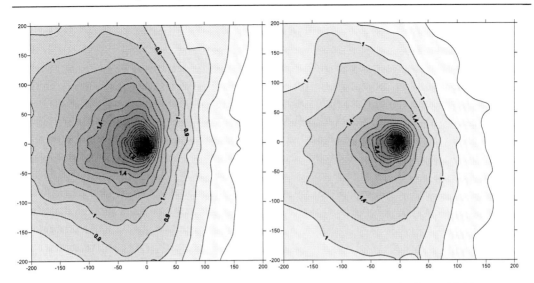

Figure 5. 95%-tile HQ for the survival of fish from simulations with (left) and without (right) wave effects (D= 0.356 m, Q = 0.11m3/s).

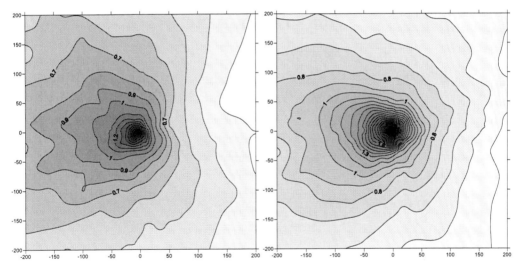

Figure 6. 95%-tile HQ for the survival of fish from simulations with (left) and without (right) wave effects (D= 0.3 m, Q = 0.11m3/s).

By comparing Figure 3 and Figure 4, it can be seen that with same maximum discharge rate (Q =0.35 m^3/s), the use of a smaller pipe (D = 0.3 m) can reduce the ecological risk. The area of concern was reduced from 7850 m^2 to about 50 m^2 without consideration of wave effects. The overall risk for the flow rate of 0.35 m^3/s is very low.

If the early stage flow rate of 0.11 m^3/s is used, the areas of concern increase (see Figure 5 and Figure 6). For the same pipe size (0.356 m), even with wave effects, the impact area with HQ greater than unity is about 70700 m^2 (radius of ~300 m). If the wave effects are ignored, the area of concern with HQ greater than unity is about 101800 m^2 (radius of ~360 m). Again, for this discharge rate, the use of a smaller pipe (D = 0.3 m) greatly reduces the area of concern. The impact area with HQ greater than unity is about 9500 m^2 (radius of ~110 m) with wave effects and about 17700 m^2 (radius of ~150 m) without.

Conclusion

The effects of waves on ecological risks of produced water have been studied in this paper. Wave effects on four different discharge scenarios were studied for the conditions similar to those at the White Rose field offshore Canada.

A probabilistic based buoyant jet spreading model was discussed. This model extends the work of Mukhtasor (2001) to include the effect of waves on initial dilution. The initial dilution model was also replaced by the Roberts *et al.* (1997) model to account for negative buoyancy.

The ecological risk was calculated by integrating the model with risk analysis software, @RISK, and conducting a Monte Carlo simulation. The simulation results showed that waves can affect the prediction of ecological risk significantly, a result that should be incorporated at the risk assessment stage. The study also showed that the risk profile is sensitive to discharge design. This offers a means to reduce ecological risks by careful design of the discharge configuration, an approach that offers the potential of inherently lower environmental impact.

Acknowledgment

Financial support from the Natural Sciences and Engineering Research Council of Canada and Petroleum Research Atlantic Canada through a Collaborative Research and Development grant (NSERC/PRAC CRD) and the Panel for Energy Research and Development (PERD) is gratefully acknowledged.

References

Bumgardner, J., Malone, C., Walker, L.F., and Shanks, R.F. (1993). Using of Monte Carlo techniques to assess POTW to compliance with EPA water quality criteria for heavy metals. *Water Environment Research*, 65(5), 674-678.

Chin, D. A. (1987). Influence of surface waves on outfall dilution. *Journal of Hydraulic Engineering*, ASCE, 113(8), 1006-1018.

Chyan, J.M., Hwung, H.H., Chang, C.Y. and Chen, I.P. (2002). Effects of discharge angles on the dilution of buoyant jet in wave motions. *Proceedings of the Fifth International Conference on Hydrodynamics*, Oct 31 – Nov 2, 2002, Tainan, Taiwan.

Domigian, A., and Waggy, W.H. (1974). Simulation – a tool for water resource management, *Water Resources Bulletin*, 10(2), 229-244.

Hodgins, D. O. and Hodgins, S. L. M. (2000). Modelled predictions of well cuttings deposition and produced water dispersion for the proposed white rose development. Seaconsult Marine Research Ltd., Vancouver, Canada.

Huang, H., Fergen, R. E., Proni, J. R. and Tsai, J.J. (1996). Probabilistic analysis of ocean outfall mixing zones. *Journal of Environmental Engineering*, ASCE, 122(5), 359-367.

Hwung, H.H., Chyan, J.M., Chang, C.Y. and Chen, Y.F. (1994). The dilution processes of alternative horizontal buoyant jets in wave motion. *Proceedings of the Twenty-fourth International Coastal Engineering Conference*, Oct 23-38, 1994, Kobe, Japan.

Meinhold, A.F., DePhillips, M.P., Holtzman, S. (1996). Final report: risk assessment for produced water discharges Louisiana open bays. Brookhaven National Laboratory, BNL-63331.

Mukhtasor (2001). Hydrodynamic modeling and ecological risk-based design of produced water discharge from an offshore platform. Ph. D. thesis, Memorial University of Newfoundland, Canada.

Okubo, A. (1971). Ocean diffusion diagrams. *Deep Sea Research*, 18, 789-802.

Roberts, P. J. W., Ferrier, Adrian, and Daviero, G. (1997). Mixing in inclined Dense Jets. *Journal of Hydraulic Engineering*, ASCE, 123(8), 693-699.

Roberts, P. J. W., and Toms, G. (1987). Inclined Dense Jets in Flowing Current. *Journal of Hydraulic Engineering*, ASCE, Vol. 113, No. 3, pp.323-341.

Tate, P. M. (2002). The rise and dilution of buoyant jets and their behavior in an internal wave field. Ph.D. thesis. University of New South Wales, Australia.

Zeitoun, M. A., Reid, R. O., Mchilhenry, W. F., and Mitchell, T. M. (1972). Model studies of outfall systems for desalination plants, Part III, Numerical simulations and design considerations. Research and Development Progress Report, Report No. 804, Office of Saline Water, U.S. Department of Interior, Washington D.C.

In: New Developments in Sustainable Petroleum Engineering
Editor: Rafiq Islam

ISBN: 978-1-61324-159-2
© 2012 Nova Science Publishers, Inc.

Chapter 2

A COMPREHENSIVE MATERIAL BALANCE EQUATION WITH THE INCLUSION OF MEMORY DURING ROCK-FLUID DEFORMATION

M. E. Hossain and M. R. Islam[*]

Department of Civil and Resource Engineering
Dalhousie University
1360 Barrington Street, Halifax, Canada

ABSTRACT

With the advent of fast computational techniques, it is time to include all salient features of the material balance equation (MBE). The inclusion of time dependent porosity and permeability can enhance the quality of oil recovery predictions to a great extent. This alteration occurs due to change of pressure and temperature of the reservoir which causes a continuous change of rock-fluid properties with time. However, few studies report such alterations and their consequences. This study investigates the effects of permeability, pore volume, and porosity with time during the production of oil. Moreover, a comprehensive MBE is presented. The equation contains a stress-strain formulation that is applicable to both rock and fluid. In addition, this formulation includes memory effect of fluids in terms of a continuous time function. Similar time dependence is also invoked to the rock strain-stress relationship. This formulation results in a highly nonlinear MBE, with a number of coefficients that are inherently nonlinear (mainly due to continuous dependence on time). This enhances the applicability of the model to fractured formations with dynamic features. Such formulation is different from previous approach that added a transient term to a steady state equation. For selected cases, the MBE is solved numerically with a newly developed nonlinear solver. A comparative study is presented using field data. Results are compared with conventional MBE approach. This comparison highlights the improvement in the recovery factor (RF) as much as 5%. This new version of MBE is applicable in the cases where rock/fluid compressibilities are available as a function of pressure from laboratory measurements or correlations. It is also applicable where non-pay zones are active as connate water and

---

[*] Corresponding author: Email: rafiqul.islam@dal.ca

solid rock expansions are strong enough with pressure depletion. Finally, suitability of the new formulation is shown for a wide range of applications in petroleum reservoir engineering.

Keywords: petroleum reservoir engineering; material balance; nonlinear material balance equation; rock-fluid deformation; stress-strain formulation; rock properties; fluid properties; memory; oil recovery.

1. Introduction

The MBE is the most fundamental equation that is used predicting petroleum reservoir performance. However, it is well known that the material balance equation that is actually solved is not a comprehensive one. The commonly used form of MBE has a number of assumptions that are not always justified. In the past, such assumptions were necessary due to limitations in computational techniques. Now a day, a high power modern computing technique is minimising the challenges of a very high accuracy and efficiency in complex calculations in reservoir simulation. Therefore, it is not necessary to have approximate solutions in reservoir engineering formulations. The MBE, one of the most widely used techniques in reservoir engineering, is an excellent examples of this. MBE is used for estimating the original hydrocarbon-in-place. It is also used for calculating the decline in average reservoir pressure with depletion. In the majority of cases, the conventional formulation of the material balance is satisfactory. However, certain circumstances, which are sometimes unpredictable, demand formulations with greater accuracy. Proper understanding of reservoir behaviour and predicting future performance are necessary to have knowledge of the driving mechanisms that control the behaviour of fluids within reservoirs. The overall performance of oil reservoir is mainly determined by the nature of energy (i.e. driving mechanism) available for moving oil toward the wellbore. There are basically six driving mechanisms that provide the natural energy necessary for oil recovery. These mechanisms are rock and liquid expansion drive, depletion drive, gas cap drive, water drive, gravity drainage drive and combination drive [Dake, 1978; Ahmed, 2002].

A lot of research works have been conducted for the last 50 years (Havlena and Odeh, 1963; Havlena and Odeh, 1964; Ramagost and Farshad, 1981; Fetkovich et al., 1991; Fetkovich et al., 1998; Rahman et al., 2006a). All these former researchers use the expansion drive mechanism in developing MBE for gas reservoir. In this study, time dependent rock/fluid properties (i.e. expressed in terms of memory function), expansion of oil, water, rock, and dissolved gas in oil and water are incorporated. In addition, total water associated with the oil reservoir volume is taken care. "Associated" water comprises connate water, water within interbedded shales and nonpay reservoir rock, and any limited aquifer volume [Fetkovich et al., 1998]. This study is an attempt to investigate the effects of expansion drive mechanism in a typical oil field using the PVT data available in the literature. Two new, rigorous MBE for oil flow in a compressible formation with residual fluid saturations are presented. One MBE is numerically solved. A new dimensionless parameter, C_{epm} in MBE is identified to represent the whole expansion drive mechanism. It is also explained how this parameter can predict the behaviour of MBE. All water and rock volumes associated with the reservoir and available for expansion, including a limited aquifer volume with formation fluid and rock volume expansion are added in this dimensionless parameter.

2. NEW COMPREHENSIVE MBE MODEL DEVELOPMENTS

2.1. New MBE for a Compressible Undersaturated Oil Reservoir

To develop a MBE for an undersaturated reservoir with no gascap gas, the reservoir pore is considered as an idealised container. MBE is derived by considering the whole reservoir as a homogeneous tank of uniform rock and fluid properties. We are considering oil and water as the only mobile phases in the compressible rock and the residual fluid saturation ($c_s \neq 0$, $c_o \neq 0$, $c_w \neq 0$, $S_{oi} \neq 0$, $S_{gi} \neq 0$, $S_{wi} \neq 0$). The derivation includes pressure-dependent rock and water compressibilities (with gas evolving solution). The inclusions of a limited aquifer volume, all water and rock volumes associated with the reservoir available for expansion are recognised. The volumetric balance expressions can be derived to account for all volumetric changes which occur during the natural productive life of the reservoir. Therefore, MBE can be presented in terms of volume changes in reservoir barrels (rb) where all the probable fluids and media changes are taken care as

Pore volume occupied by the oil initially in place + originally dissolved gas at time, t = 0 and p_i, rb
=
Pore volume occupied by the remaining oil at a given time, t and at p, rb
+
Change in oil volume due to oil expansion at a given time t and at p, $(-\Delta V_o)$, rb
+
Change in water volume due to connate water expansion at a given time t and at p, $(-\Delta V_w)$, rb
+
Change in dissolved gas volume due to gas expansion at a given time, t and at p, $(-\Delta V_g)$, rb
+
Change in pore volume due to reduction a given time, t and at p, (ΔV_s), rb
+
Change in associated volume due to expansion and reduction of water and pore volume at a given time, t and at p, (ΔV_A), rb
+
Change in water volume due to water influx and water production at a given time, t and at p, rb

(1)

where:

ΔV_A = change in associated volume due to expansion and reduction of water and pore volume at a given time, t and at p, (ΔV_A), rb
ΔV_g = change in gas volume due to oil expansion at a given time t and at p, $V_{gi} - V_g$, rb
ΔV_o = change in oil volume due to oil expansion at a given time t and at p, $V_{oi} - V_o$, rb
ΔV_s = change in pore volume due to rock contraction at a given time t and at p, $V_{si} - V_s$, rb
ΔV_w = change in water volume due to oil expansion at a given time t and at p, $V_{wi} - V_w$, rb
V_g = gas volume at a reduced pressure p, rb
V_o = oil volume at a reduced pressure p, rb
V_s = solid rock pore volume at a reduced pressure p, rb
V_w = water volume at a reduced pressure p, rb
V_{gi} = gas volume at initial pressure p_i, rb
V_{oi} = oil volume at initial pressure p_i, rb
V_{si} = solid rock pore volume at initial pressure p_i, rb
V_{wi} = water volume at initial pressure p_i, rb

Now

1) Pore volume occupied by the initial oil in place and originally dissolved gas = NB_{oi}
2) Pore volume occupied by the remaining oil at p = $(N - N_p)B_o$

where

B_o = oil formation volume factor at a given time t and a reduced pressure p, rb/stb
B_{oi} = oil formation volume factor at initial pressure p_i, rb/stb
N = initial oil in place (e.g. initial volume of oil in reservoir), = $[V\phi(1 - S_{wi})/B_{oi}]$, stb
N_P = cumulative oil production at a given time t and a reduced pressure, p, stb
V = total reservoir volume, ft^3
ϕ = variable rock porosity with space and time, volume fraction
S_{wi} = water saturation at initial pressure p_i

The isothermal fluid and formation compressibilities are defined according to the above discussion as [Dake, 1978; Ahmed, 2000; Rahman et al., 2006a]:

Oil:
$$c_o = -\frac{1}{V_o}\frac{\partial V_o}{\partial p}\bigg]_T \qquad (2)$$

Water:
$$c_w = -\frac{1}{V_w}\frac{\partial V_w}{\partial p}\bigg]_T \qquad (3)$$

Gas:
$$c_g = -\frac{1}{V_g}\frac{\partial V_g}{\partial p}\bigg]_T \qquad (4)$$

Solid rock formation:

$$c_s = \frac{1}{V_s}\frac{\partial V_s}{\partial p}\bigg]_T \tag{5}$$

If p_i is the initial reservoir pressure and p is the average reservoir pressure at current time t, one can write down the expressions for ΔV_o, ΔV_w, ΔV_g, and ΔV_s by integrating Equation (2) through Equation (5), assuming compressibilities to be pressure dependent, and by subsequent algebraic manipulations as (Rahman et al., 2006a):

$$\Delta V_o = -V_{oi}\left(1 - e^{\int_p^{p_i} c_o\, dp}\right) \tag{6}$$

$$\Delta V_w = -V_{wi}\left(1 - e^{\int_p^{p_i} c_w\, dp}\right) \tag{7}$$

$$\Delta V_g = -V_{gi}\left(1 - e^{\int_p^{p_i} c_g\, dp}\right) \tag{8}$$

$$\Delta V_s = V_{si}\left(1 - e^{-\int_p^{p_i} c_s\, dp}\right) \tag{9}$$

where,

c_g = reservoir gas compressibility at a reduced pressure p, psi^{-1}
c_o = reservoir oil compressibility at a reduced pressure p, psi^{-1}
c_s = reservoir rock formation compressibility at a reduced pressure p, psi^{-1}
c_w = reservoir water compressibility at a reduced pressure p, psi^{-1}
p = current reservoir pressure (at time t), $psia$
p_i = initial reservoir pressure, $psia$

In addition to above, associated volume is also included in developing a MBE with expansion drive mechanism. The "associated" volume is an additional reservoir part which is not active in oil/gas production. However, this part of the reservoir may accelerate the oil recovery by its water and rock expansion-contraction. Fetkovich et al. (Fetkovich et al., 1991; Fetkovich et al., 1998) defined this term as "the nonnet pay part of reservoir where interbedded shales and poor quality rock is assumed to be 100% water-saturated". The interbedded nonnet pay volume and limited aquifer volumes are referred to as "associated" water volumes and both contribute to water influx during depletion. They used a volume fraction, M to represent the associated volume effects in the conventional MBE (Dake, 1978; Ahmed, 2000; Craft and Hawkins, 1959; Havlena and Odeh, 1963; Havlena and Odeh, 1964) for gas reservoir. M is defined as the ratio of associated pore volume to reservoir pore volume. If we consider this associated volume change, this part will be contributing as an additional expansion term in MBE. Therefore, the volume change would be:

$$\Delta V_A = -(-\Delta V_{Aw}) + \Delta V_{As} = V_{Awi}\left(e^{\int_p^{p_i} c_w\, dp} - 1\right) + V_{Asi}\left(1 - e^{-\int_p^{p_i} c_s\, dp}\right) \tag{10}$$

where

ΔV_{As} = change in associated volume due to rock expansion at a given time t and at p, $V_{Asi} - V_{As}$, rb
ΔV_{Aw} = change in associated volume due to water expansion at a given time t and at p, $V_{Awi} - V_{Aw}$, rb
V_{Asi} = associated solid rock pore volume at initial pressure p_i, rb
V_{Awi} = associated water volume at initial pressure p_i, rb

Note that ΔV_o, ΔV_w, and ΔV_g have negative values due to expansion, and ΔV_s has a positive value due to contraction. The initial fluid and pore volumes can be expressed as

$$V_{oi} = \frac{NB_{oi}}{1-S_{wi}} S_{oi} \tag{11}$$

$$V_{wi} = \frac{NB_{oi}}{1-S_{wi}} S_{wi} \tag{12}$$

$$V_{gi} = \frac{(NB_{oi}) \times (R_{soi}/B_{oi}) \times B_{gi} + (NB_{oi}) \times (R_{swi}/B_{wi}) \times B_{gi}}{1-S_{wi}} S_{gi} \tag{13}$$

$$V_{si} = \frac{NB_{oi}}{1-S_{wi}} \tag{14}$$

$$V_{Awi} = M \frac{NB_{oi}}{1-S_{wi}} \text{ (as 100\% water saturated e.g., } S_w = 1.0) \tag{15}$$

$$V_{Asi} = M \frac{NB_{oi}}{1-S_{wi}} \tag{16}$$

where

B_{wi} = water formation volume factor at initial pressure p_i, rb/stb
B_{gi} = gas formation volume factor at initial pressure p_i, rb/scf
S_{wi} = water saturation at initial pressure p_i
S_{gi} = gas saturation at initial pressure p_i
R_{si} or R_{soi} = gas solubility at initial reservoir pressure e.g. initial solution gas-oil ratio, scf/stb
R_{swi} = initial solution gas-water ratio, scf/stb

Substituting Equations (11) through (16) in Equations (6) to (10) respectively, one can write down the equations as:

$$\Delta V_o = -V_{oi}\left(1 - e^{\int_p^{p_i} c_o \, dp}\right) = -\frac{NB_{oi}}{1-S_{wi}} S_{oi}\left(1 - e^{\int_p^{p_i} c_o \, dp}\right) \tag{17}$$

$$\Delta V_w = -V_{wi}\left(1 - e^{\int_p^{p_i} c_w \, dp}\right) = -\frac{NB_{oi}}{1-S_{wi}} S_{wi}\left(1 - e^{\int_p^{p_i} c_w \, dp}\right) \tag{18}$$

$$\Delta V_g = -V_{gi}\left(1 - e^{\int_p^{p_i} c_g\, dp}\right)$$
$$= -\frac{(NB_{oi}) \times (R_{soi}/B_{oi}) \times B_{gi} + (NB_{oi}) \times (R_{swi}/B_{wi}) \times B_{gi}}{1 - S_{wi}} S_{gi}\left(1 - e^{\int_p^{p_i} c_g\, dp}\right)$$

(19)

$$\Delta V_s = V_{si}\left(1 - e^{-\int_p^{p_i} c_s\, dp}\right) = \frac{NB_{oi}}{1 - S_{wi}}\left(1 - e^{-\int_p^{p_i} c_s\, dp}\right) \quad (20)$$

$$\Delta V_A = \Delta V_{Aw} + \Delta V_{As} = M\frac{NB_{oi}}{1-S_{wi}}\left(e^{\int_p^{p_i} c_w\, dp} - 1\right) + M\frac{NB_{oi}}{1-S_{wi}}\left(1 - e^{-\int_p^{p_i} c_s\, dp}\right) \quad (21)$$

Substituting Equation (17) through (21) in Equation (1), one can write down the MBE as:

$$N_p B_o - (W_e - W_p B_w) = N(B_o - B_{oi} + B_{oi} C_{epm}) \quad (22)$$

where

$$C_{epm} = \left\{\frac{\begin{array}{c}S_{oi}\left(e^{\int_p^{p_i} c_o\, dp}-1\right)+S_{wi}\left(e^{\int_p^{p_i} c_w\, dp}-1\right)+S_{gi}\left(e^{\int_p^{p_i} c_g\, dp}-1\right)\left(\frac{R_{soi}}{B_{oi}}+\frac{R_{swi}}{B_{wi}}\right)B_{gi}\\ +\left(1-e^{-\int_p^{p_i} c_s\, dp}\right)+M\left[\left(e^{\int_p^{p_i} c_w\, dp}-1\right)+\left(1-e^{-\int_p^{p_i} c_s\, dp}\right)\right]\end{array}}{1-S_{wi}}\right\} \quad (23)$$

where

B_w = water formation volume factor at a given time t and a reduced pressure p, rb/stb
W_p = cumulative water production at p, stb
W_e = cumulative water influx into reservoir at p, rb
C_{epm} = parameter of effective compressibility due to residual fluid, dissolved gas and formation for the proposed MBE, dimensionless

Equation (22) is the new, rigorous and comprehensive MBE for an undersaturated reservoir with no gascap gas and above the bubble-point pressure. The above MBE is very rigorous because it considers the fluid and formation compressibilities as any functions of pressure. It is deeming oil and water as the only mobile phases in the compressible rock. This rigorous MBE is applicable for a water-drive system with a history of water production in an undersaturated reservoir. Here, associated volume ratio, the residual and dissolved phase saturations are also considered. The Equation (23) is an expression of the proposed dimensionless parameter, C_{epm} where all the probable and available expansions are illustrated. When the water influx term is not significant [$i.e.$, $W_e = 0$], Equation (22) may be modified to:

$$N_p B_o + W_p B_w = N(B_o - B_{oi} + B_{oi} C_{epm}) \tag{24}$$

Equation (24) can be written in the form of straight line's MBE as (Havlena and Odeh, 1963; Havlena and Odeh, 1964):

$$N = \frac{F}{E_0 + E_{cepm}} \tag{25}$$

where

$$F = N_p B_o + W_p B_w$$
$$E_0 = B_o - B_{oi}$$
$$E_{cepm} = B_{oi} C_{epm}$$

Now, if we consider a constant data of oil, water, gas and formation compressibilities (*e.g.*, compressibilities are not functions of pressure), Equation (22) remains unchanged. However, Equation (23) can be modified by integrating the power of exponents as:

$$C_{epm} = \left\{ \frac{S_{oi}\left(e^{c_o(p_i-p)} - 1\right) + S_{wi}\left(e^{c_w(p_i-p)} - 1\right) + S_{gi}\left(e^{c_g(p_i-p)} - 1\right)\left(\frac{R_{soi}}{B_{oi}} + \frac{R_{swi}}{B_{wi}}\right)B_{gi} + \left(1 - e^{-c_s(p_i-p)}\right) + M\left[\left(e^{c_w(p_i-p)} - 1\right) + \left(1 - e^{-c_s(p_i-p)}\right)\right]}{1 - S_{wi}} \right\} \tag{26}$$

Equation (26) is still a rigorous expression for use in the case of constant compressibilities. This equation can be further approximated by the exponential terms for small values of the exponents as:

$$e^{c_o(p_i-p)} \approx 1 + c_o(p_i - p) \tag{27}$$

$$e^{c_w(p_i-p)} \approx 1 + c_w(p_i - p) \tag{28}$$

$$e^{c_g(p_i-p)} \approx 1 + c_g(p_i - p) \tag{29}$$

$$e^{-c_s(p_i-p)} \approx 1 - c_s(p_i - p) \tag{30}$$

Substituting Equations (27) through (30) in Equation (26) and considering the average reservoir pressure, it becomes:

$$C_{epm} = \frac{S_{oi} c_o + S_{wi} c_w + S_{gi} c_g \left(\frac{R_{soi}}{B_{oi}} + \frac{R_{swi}}{B_{wi}}\right) B_{gi} + c_s + M(c_w + c_s)}{1 - S_{wi}} \Delta p \tag{31}$$

where:

$\bar{p}$ = volume average reservoir pressure at p, psi

Δp = average pressure drop in the reservoir at p, $= p_i - \bar{p}$, psi

The dimensionless parameter, C_{epm} expressed in terms of fluids (oil, water and gas) and formation compressibilities, and saturation is same to that of conventional MBE if we neglect the expansion of dissolved gas in oil and water, and the associated volume expansion. This issue will be discussed later.

2.2. Conventional MBE

The general form of conventional MBE can be written as [Dake, 1978; Ahmed, 2000; Craft and Hawkins, 1959; Havlena and Odeh, 1963; Havlena and Odeh, 1964]:

$$N_p B_o + W_p B_w = N(B_o - B_{oi} + B_{oi} C_{eHO}) \tag{32}$$

where

$$C_{eHO} = \frac{S_{oi} c_o + S_{wi} c_w + c_s}{1 - S_{wi}} \Delta p \tag{33}$$

C_{eHO} = parameter of effective compressibility due to residual water and formation for the Havlena and Odeh [1963] MBE, dimensionless

Equation (24) resembles the MBE (Equation (32)) of Havlena and Odeh (1963, 1964) for a volumetric and undersaturated reservoir with no gascap except the pattern of dimensionless parameter, C_{epm} and Δp. Equation (32) can further be written as in the straight line form of Havlena and Odeh MBE as:

$$N = \frac{F}{E_0 + E_{ceHO}} \tag{34}$$

where

$$F = N_p B_o + W_p B_w$$
$$E_0 = B_o - B_{oi}$$
$$E_{ceHO} = B_{oi} C_{eHO}$$

Now, if we neglect dissolved gas saturation and associated volume expansion in Equation (31) and define the pressure at time t as the average pressure, the equation becomes as:

$$C_{epm} = \frac{S_{oi} c_o + S_{wi} c_w + c_s}{1 - S_{wi}} \Delta p \tag{35}$$

The right hand side of above equation is same as stated in Equation (33).

2.3. A Comprehensive MBE with Memory for Cumulative Oil Recovery

Equation (31) can be written using variable pressure expression as:

$$C'_{epm} = \frac{\left[S_{oi}c_o + S_{wi}c_w + S_{gi}c_g\left(\frac{R_{soi}}{B_{oi}} + \frac{R_{swi}}{B_{wi}}\right)B_{gi} + c_s + M(c_w + c_s)\right] \times [p_i - p_{(t)}]}{1 - S_{wi}} \qquad (36)$$

In Equation (36), the average pressure decline for a particular time t from the start of production may be calculated using the time-dependent rock/fluid properties with stress-strain model. The mathematical explanation and the derivation of the stress-strain formulation are described in Hossain et al. (2007a). They gave the stress-strain relationship as follows:

$$\tau_T = (-1)^{0.5} \times \left(\frac{\partial \sigma}{\partial T}\frac{\Delta T}{\alpha_D M_a}\right) \times \left[\frac{\int_0^t (t-\xi)^{-\alpha}\left(\frac{\partial^2 p}{\partial \xi \partial x}\right)d\xi}{\Gamma(1-\alpha)}\right]^{0.5} \times \left(\frac{6K\mu_0\eta}{\frac{\partial p}{\partial x}}\right)^{0.5} \times e^{\left(\frac{E}{RT_T}\right)}\frac{du_x}{dy} \qquad (37)$$

where:

$E=$ activation energy for viscous flow, Btu/mol
$K=$ operational parameter
$M_a =$ Marangoni number
$R=$ universal gas constant, $Btu/mole - °F$
$t=$ time, s
$u=$ filtration velocity in x direction, ft/s
$u_x=$ fluid velocity in porous media in the direction of x axis, m/s
$T_T =$ temperature of the reservoir at time, t, $°F$
$T_o =$ initial reservoir temperature at time, $t = 0$, $°F$
$\frac{du_x}{dy}=$ velocity gradient along y-direction, $1/s$
$\left|\frac{d\sigma}{dT}\right|=$ the derivative of surface tension σ with temperature and can be positive or negative depending on the substance, $lb_f/ft - °F$
$\Delta T = T_T - T_0 =$ temperature difference, $°F$
$\beta=$ coefficient of thermal expansion, $1/°F$
$\sigma=$ surface tension, lb_f/ft
$\alpha=$ fractional order of differentiation, dimensionless
$\alpha_D=$ thermal diffusivity, ft^2/s
$\rho_f=$ density of fluid, lb_m/ft^3
$\rho_r=$ a density of fluid, lb_m/ft^3
$\mu =$ dynamic viscosity of reservoir fluid at temperature T, cp
$\mu_0=$ fluid dynamic viscosity at initial reservoir temperature T_0, pa-s
$\tau_T=$ shear stress at temperature T, Pa
$\xi=$ a dummy variable for time i.e. real part in the plane of the integral, s
$\eta=$ ratio of the pseudopermeability of the medium with memory to fluid viscosity, $ft^3 s^{1+\alpha}/lb_m$

The above equation reduces to (Hossain, 2008):

$$p_i - p_{(t)} = -\frac{6K\mu_0\eta\left(\frac{\Delta T}{a_D M_a}\frac{\partial\sigma}{\partial T}\right)^2 \times \left[\frac{\int_0^t(t-\xi)^{-\alpha}\left(\frac{\partial^2 p}{\partial\xi\partial x}\right)d\xi}{\Gamma(1-\alpha)}\right] \times e^{2\left(\frac{E}{RT_T}\right)} \times \left(\frac{du_x}{dy}\right)^2}{\tau_T^2} u_x \Delta t \tag{38}$$

The change of pressure with time and space can be calculated using the stress-strain Equation (38). This change of pressure is directly related to oil production performance of a well. Therefore, substituting Equation (38) into Equation (36):

$$C'_{epm} = \frac{\left[S_{oi}c_o + S_{wi}c_w + S_{gi}c_g\left(\frac{R_{soi}}{B_{oi}} + \frac{R_{swi}}{B_{wi}}\right)B_{gi} + c_s + M(c_w + c_s)\right] \times \left[-\frac{6K\mu_0\eta(L-x)\left(\frac{\partial\sigma}{\partial T}\frac{\Delta T}{a_D M_a}\right)^2 \left[\int_0^t(t-\xi)^{-\alpha}\left(\frac{\partial^2 p}{\partial\xi\partial x}\right)d\xi\right]e^{2\left(\frac{E}{RT_T}\right)}\left(\frac{du_x}{dy}\right)^2}{\tau_T^2\{\Gamma(1-\alpha)\}}u_x\Delta t\right]}{1 - S_{wi}} \tag{39}$$

where C'_{epm} is the modified dimensionless parameter which depends on rock/fluid memory and other related fluid and rock properties and Δt is the time difference between the start of production and a particular time which is actually time, t.

The C'_{epm} of Equation (39) can be used in Equation (22) to represent the time dependent rock/fluid properties and other properties which are related to the formation fluid and formation itself. Therefore, Equation (22) can be written as (Hossain, 2008):

$$N_p = \phi A_3 + \frac{1}{B_o}(W_e - W_p B_w) \tag{40}$$

where

$$N = \frac{V\phi(1 - S_{wi})}{B_{oi}}$$

$$A_3 = \frac{V(1 - S_{wi})}{B_{oi} B_o}(B_o - B_{oi} + B_{oi} C'_{epm})$$

The Equation (40) represents a new rigorous MBA with memory where the every possibility of time dependent rock/fluid properties is considered.

3. SIGNIFICANCE OF C_{epm}

The dimensionless parameter, C_{epm}, in the below Equation (41) can be considered as the effective strength of the energy source for oil production in expansion drive oil recovery. This is only due to the compressible residual fluids (oil, water and dissolved gas) and rock expansions of the reservoir. This value does not account for the oil compressibility as stated by other researchers [Dake, 1978; Fetkovich et al., 1991; Fetkovich et al., 1998; Ahmed, 2000; Rahman et al., 2006b]. If we see the expression of C_{epm} as presented in Equation (23), it is a function of the current reservoir pressure, fluid compressibilities, initial saturations, dissolved gas properties engaged in water and oil, and associated volume fraction. C_{epm} in Equation (26) is still a function all those parameters except a set of constant compressibilities instead of variable compressibilities. The final simplified form of C_{epm} presented in Equation (31) is dependent on current average reservoir pressure drop and other related parameters as stated above. The dimensionless parameter, C_{epm} is an important parameter in the proposed MBE (Equation (22)) because it can be used as an analytic tool to predict how the MBE will behave for the relevant input data. An elaborate discussion is stated in results and discussion.

Some specific significant properties may be well explained to elucidate the new version of C_{epm}, presented in Equations (23) and (26). To explain these significance of C_{epm}, Equation (22) can be rearranged as:

$$\frac{1}{B_o}\left(1 - C_{epm} - \frac{W_e - W_p B_w}{N B_{oi}}\right) = \frac{1}{B_{oi}}\left(1 - \frac{N_p}{N}\right) \tag{41}$$

The Equation (41) is used to explain the effects of C_{epm}, where two cases have been considered:

3.1. Water Drive Mechanism with Water Production

The Equation (41) gives a limit of expansion plus water drives mechanism for initial fluid situations, water influx and water production. This limit may be expressed as:

$$0 \le \left(C_{epm} - \frac{W_e - W_p B_w}{N B_{oi}}\right) < 1.0 \tag{42}$$

The Equation (42) is true for any given average reservoir pressure. The lower limit in Equation (42) is due to the fact of C_{epm}, W_e and W_p where all these parameters zero at the initial reservoir pressure. The upper limit is characterised by the fact that the right-hand side term in Equation (41) is zero, when all the original oil-in-place has been produced. However, practically it is not possible to reach the production level up to that marks. Therefore, the upper limit should be less than 1. Hence, it may be concluded that if the numerical values beyond these range comes out, there might be some problem in input data or there might be a problem in calculating or assigning the average reservoir pressure. So, it is a tool to diagnose or predict the reservoir behaviour in the early stage of production.

3.2. Depletion Drive Mechanism with No Water Production

When depletion drive mechanism (with no water influx) with no water production is considered, Equation (41) gives a limit for this drive mechanism as:

$$0 \le C_{epm} < 1.0 \tag{43}$$

The limits are identified using the same argument as the water drive mechanism, presented in the previous section. If the cited limits are violated at a given time, there is no chance of calculating any reasonable values of the average reservoir pressure. Therefore, the limits are the indications of decision tool about the reservoir and fluid properties and decline criteria.

4. NUMERICAL SIMULATION

The numerical results of the dimensionless parameters, C_{epm} and C_{eHO} based on the models presented by Equations (23), (26), (31), and (33) can be obtained by solving these equations. A volumetric undersaturated reservoir with no gascap gas is considered for the simulation. The reservoir initial pressure is $p_i = 4000\ psi$.

Table 1. Reservoir rock and fluid properties for simulation

Rock and fluid properties [Hall, 1953; Dake, 1978; Ahmed, 2000]	
B_{gi} = 0.00087 rb/scf	c_w = 3.62 x $10^{-6}\ psi^{-1}$
B_{oi} = 1.2417 rb/stb	R_{soi} = 510.0 scf/stb
B_{wi} = 1.0 rb/stb	R_{swi} = 67.5 scf/stb
c_g = 500.0 x $10^{-6}\ psi^{-1}$	$S_{gi} = 20\%$
c_o = 15.0 x $10^{-6}\ psi^{-1}$	$S_{oi} = 60\%$
c_s = 4.95 x $10^{-6}\ psi^{-1}$	$S_{wi} = 20\%$

Table 1 presents the rock and fluid properties that have been used in solving the above mentioned equations. Trapezoidal method is used to solve the exponential integral. All computation is carried out by Matlab 6.5. To calculate IOIP using Havlena and Odeh (1963, 1964) straight line method, the Virginia Hills Beaverhill Lake field [Ahmed, 2002] data and an additional data of Table 1 are considered. The initial reservoir pressure is 3685 psi. The bubble-point pressure was calculated as 1500 psi. Table 2 shows the field production and PVT data.

Table 2. The field production and PVT data (Example 11-3: of Ahmed, 2002)

Volumetric average pressure psi	No. of producing wells	B_o rb/stb	N_p mstb	W_p mstb
3685	1	1.3102	0	0
3680	2	1.3104	20.481	0
3676	2	1.3104	34.750	0
3667	3	1.3105	78.557	0
3664	4	1.3105	101.846	0
3640	19	1.3109	215.681	0
3605	25	1.3116	364.613	0
3567	36	1.3122	542.985	0.159
3515	48	1.3128	841.591	0.805
3448	59	1.3130	1273.530	2.579
3360	59	1.3150	1691.887	5.008
3275	61	1.3160	2127.077	6.500
3188	61	1.3170	2575.330	8.000

5. RESULTS AND DISCUSSION

5.1. Effects of Compressibilities on Dimensionless Parameters

Figures 1(a)–(d) present the variation of dimensionless parameters with average reservoir pressures when associated volume fraction does not take care. The figures give a general idea of how the fluid and formation compressibilities play a role on MBE when pressure varies. The plotting of Figure 1(a) is based on Equation (23) where variable compressibilities of the fluids and formation are taken care. The trend of the curve is a nonlinear exponential type. When reservoir pressure starts to deplete, C_{epm} increases and it reaches its highest value at $p = 0$. Figure 1(b) is plotted based on Equation (26) where constant compressibilities of the fluids and formation are considered. The trend of the curve is still in the form of a nonlinear exponential type. The numerical data and shape of the curve is almost same as Figure 1(a). Figure 1(c) is plotted using the Equation (31) where constant compressibilities and an approximation of the exponential terms are considered. A straight line curve produces where the numerical values are less than that of Figures 1(a) and 1(b). The dimensionless parameter from the use of conventional MBE (Equation 33) shows a straight line curve where the numerical values are much less than that of the previous presentation (Figure 1(d)).

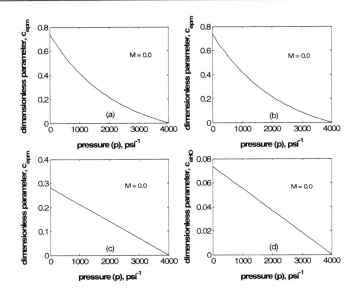

Figure 1. Dimensionless parameter variation with pressure for different equations.

5.2. Comparison of Dimensionless Parameters Based on Compressibility Factor

Figure 2 explains how the values of the dimensionless parameter can vary with reservoir pressure for a given set of compressibility and saturation.

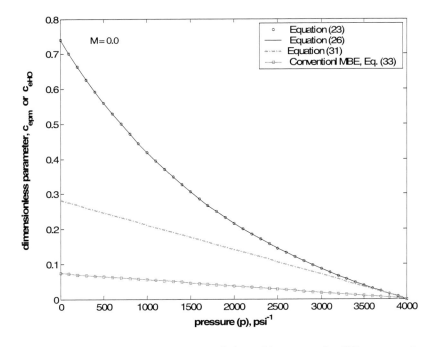

Figure 2. Comparison of dimensionless parameters variation with pressure for different equations.

It compares between the dimensionless parameter, C_{epm} of the proposed model and the conventional dimensional parameter, C_{eHO}. The curves of the figure have been generated using Equations (23), (26), (31) and (33). The depleted reservoir pressure of $p = 0$ gives the maximum value of C_{epm} or C_{eHO}. The constant or variable compressibility (Equations (23), and (26)) does not make any significant difference in computation up to a certain level of accuracy, which is approximately 10^{-3} %. At low pressures, the magnitude of C_{epm} increases very fast comparing with other two Equations (31) and (33). The pattern and nature of the curves are already explained in Figure 1. The change of dimensionless parameter is low for conventional MBE. However, when the expression of the proposed MBE's parameter is simplified, it turns to reduce the magnitude of the dimensionless parameter. This cleanly means that the simplified version of the proposed model will be the same if it will be further simplified. Therefore, it may be concluded that the use of conventional MBE over estimate the IOIP. The proposed model is closer to reality. This issue will be discussed in later section.

5.3. Effects of *M* on Dimensionless Parameter

If we consider the associated volume in the reservoir, all the available or probable pressure support from rock and water as well as from fluids are being accounted for the proposed MBE with dimensionless parameter. Figure 3 has been generated for a specific reservoir where several *M* values have been considered. The figure shows the variation of C_{epm} with average reservoir pressure for different *M* values. Figures 3(a)–(c) present the C_{epm} variation for the proposed Equations (23), (26) and (31) respectively.

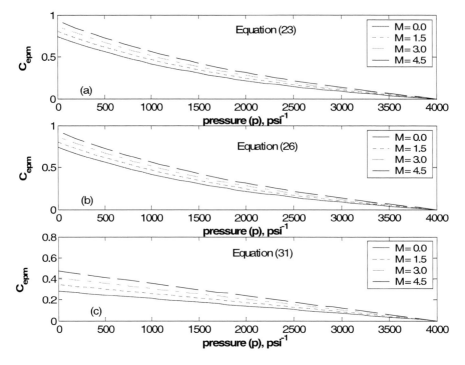

Figure 3. Dimensionless parameter variations with pressure at different *M* ratios.

These curves have specific characteristics depending on the pressure dependence of rock and fluids (water, oil and dissolved gas) compressibilities. These curves have relatively less variant at high pressure, increase gradually as pressure decreases, and finally rise sharply at low pressure especially after 1,000 psi. All the curves in Figure 3 have the same characteristics except the numerical values of the dimensionless parameter, C_{epm}. For every equation, if M increases, the curve shifts upward in the positive direction of C_{epm}. The difference in C_{epm} due to M is more dominant at low pressure. This trend of the curve indicates that the matured reservoir feels more contributions from associated volume of the reservoir.

5.4. Effects of Compressibility Factor with M Values

Figures 4(a)–(c) illustrate C_{epm} verses pressure for different proposed Equations (23), (26), and (31) at several M values of 0.0, 1.5, 3.0, 4.5 respectively. The shape and characteristics of all curves are same as Figure 3. When variable compressibilities are considered (Equation (23)) with pressure, there is a big difference at low pressure with constant compressibilities and exponential approximation Equation (Equation (31)) for all M values. However, there is no significant change in Equation (23) and (26) at different M values. It should be mentioned here that as M increases, C_{epm} increases. This is true for all the equations.

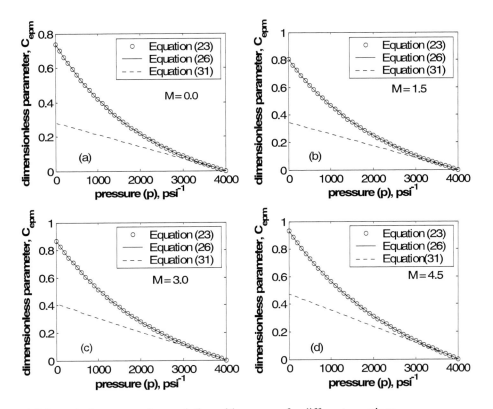

Figure 4. Dimensionless parameters variation with pressure for different equations.

5.5. Comparison of Models Based on RF

Figure 5 illustrates the underground withdrawal, F verses the expansion term $E_0 + E_{cepm}$ for the proposed MBE (Equation (25)) with Equations (26) and (31) where associated volume ratio is ignored. The conventional MBE (Equation (34)) with Equation (33) is also shown in the same graph. Using best fit curve fitting analysis, these plotting give a straight line passing through the origin with a slop of N. IOIP is identified as 73.41 mmstb, 68.54 mmstb and 175.75 mmstb for the MBE with Equations (26), (31), and (33) respectively. The corresponding recovery factors are calculated as 3.76%, 3.51%, and 1.46%. Therefore, the inclusions of the probable parameters increase the ultimate oil recovery. Here, linear plot indicates that the field is producing under volumetric performance ($W_e = 0$) which is strictly by pressure depletion and fluid and rock expansion.

Figure 6 shows a plotting of $(F/E_0 + E_{cepm})$ verses cumulative production, N_p for the proposed MBE (Equation (24)) with Equations (26) and (31), and $(F/E_0 + E_{ceHO})$ vs. N_p for the conventional MBE (Equation (32)) with Equation (33). In this figure, associated volume ratio is also ignored.

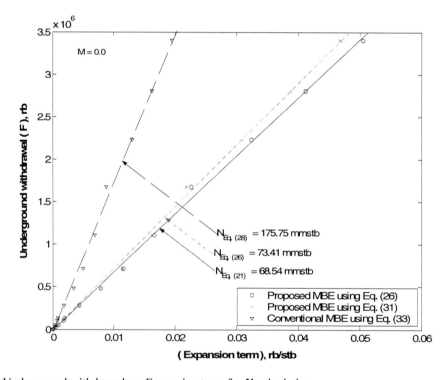

Figure 5. Underground withdrawal vs. Expansion term for N calculation.

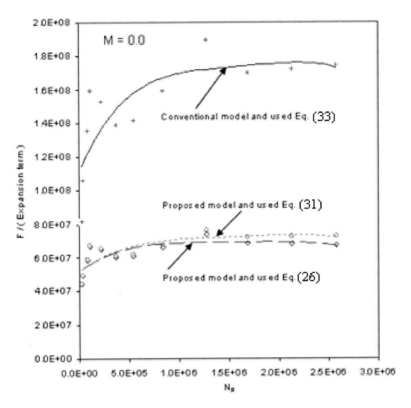

Figure 6. $F/(E_0 + E_{cepm} \text{ or } E_{ceHO})$ vs. N_P.

The best fit plot for all the equations indicate that the reservoir has been engaged by water influx, abnormal pore compaction or a combination of these two [Dake, 1978; Ahmed, 2000]. In our situation we ignore the water influx ($W_e = 0$). Therefore, we may conclude that the reservoir behaviour is an indication of pore compactions and fluids and rocks expansion.

5.6. Effects of *M* on MBE

The recovery factor of the proposed model is higher than that of conventional MBE (Figure 5). Moreover, the comprehensive proposed MBE, Equation (25) with Equation (23) has higher RF than that of using Equation (26) with Equation (25). Therefore, to show the effects of *M* values, Figure 7 illustrates F vs. $(E_0 + E_{cepm})$ for only the proposed MBE with Equation (26). The straight line plotting passing through the origin of the figure gives 59.78 mmstb, 53.01 mmstb and 47.61 mmstb of IOIP for $M = 1.0, 2.0, 3.0$ respectively. The corresponding RF values are calculated as 4.31%, 4.86%, and 5.4%. So, RF increases with the increase of *M* values which correspond that if there is an associated volume of a reservoir, it should be considered in the MBE calculations otherwise there might be some error in getting the true production history of reservoir life.

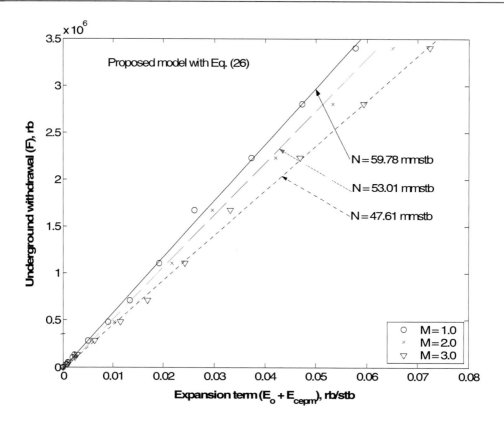

Figure 7. Underground withdrawal vs. Expansion term for different M values.

Conclusion

A new comprehensive material balance equation has been established for an undersaturated oil reservoir with no gascap gas. The proposed MBEs have the core concepts of using variable compressibilities, residual fluid saturations and time dependent rock/fluid properties. The associated volume of the reservoir is also accounted to derive the generalised MBE. The MBE is greatly influenced by the compressibilities of fluids as well as rocks which help to increase the RF values in production history. If there exists an additional reservoir part which is not active in oil production (e.g. *M* values), RF is also affected by these *M* values. All these considerations offer the unique features of the proposed MBE with improved RF. Therefore, the inclusions of all probable parameters increase the ultimate oil recovery. A general idea of how the fluid and formation compressibilities play a role on MBE can be known by using this MBE. The available literature support that MBE has a linear relationship. However, the proposed MBE is a nonlinear type which is obvious due to nonlinear nature of pressure decline with time or distance. The input data can be scanned using the dimensionless parameter, C_{epm} expression.

ACKNOWLEDGMENT

The authors would like to thank the Atlantic Canada Opportunities Agency (ACOA) for funding this project under the Atlantic Innovation Fund (AIF). The first author would also like to thank Natural Sciences and Engineering Research Council of Canada (NSERC) for funding.

REFERENCES

Ahmed, T. (2002) *Reservoir Engineering Handbook*. 2nd edition. Gulf Professional Publishing, Boston, U.S.A.

Craft, B.C. and Hawkins, M.F. (1959) *Applied Petroleum Reservoir Engineering*. Prentice-Hall, Inc., Englewood Cliffs, NJ 07632.

Dake, L.P. (1978) *Fundamentals of Reservoir Engineering*. Elsevier Science Publishing Company Inc., New York, NY 10010, U.S.A.

Fetkovich, M.J., Reese, D.E. and Whitson, C.H. (1991) Application of a General Material Balance for High-Pressure Gas Reservoir. Paper SPE 22921, presented at the *1991 SPE Annual Technical Conference and Exhibition*, Dallas, October 6-9.

Fetkovich, M.J., Reese, D.E. and Whitson, C.H. (1998) Application of a General Material Balance for High-Pressure Gas Reservoir, *SPE Journal*, (March), pp. 3-13.

Hall, H.N. (1953) Compressibility of Reservoir Rocks. *Trans. AIME*, 198, pp. 309-311.

Havlena, D. and Odeh, A.S. (1963) The Material Balance as an Equation of a Straight Line. JPT (August) 896, *Trans., AIME*, 228.

Havlena, D. and Odeh, A.S. (1964) The Material Balance as an Equation of a Straight Line-Part II, Field Cases. JPT (July) 815, *Trans., AIME*, 231.

Hossain, M.E. (2008) An Experimental and Numerical Investigation of memory-Based Complex Rheology and Rock/Fluid Interactions, PhD dissertation, Dalhousie University, Halifax, Nova Scotia, Canada, pp. 793.

Hossain, M.E., Mousavizadegan, S.H., Ketata, C. and Islam, M.R. (2007) A Novel Memory Based Stress-Strain Model for Reservoir Characterization, *Journal of Nature Science and Sustainable Technology*, Vol. 1(4), pp. 653 – 678.

Rahman, N.M.A., Anderson, D.M. and Mattar, L. (2006a) New Rigorous Material Balance Equation for Gas Flow in a Compressible Formation with Residual Fluid Saturation. SPE 100563, presented at the SPE Gas Technology Symposium held in Calgary, Alberta, Canada, May 15-17.

Rahman, N.M.A., Mattar, L. and Zaoral, K. (2006b) A New Method for Computing Pseudo-Time for Real Gas Flow Using the Material Balance Equation. *J. of Canadian Petroleum Technology*, 45(10), pp. 36-44.

Ramagost, B.P. and Farshad, F.F. (1981) P/Z Abnormally Pressured Gas Reservoirs, paper *SPE 10125 presented at SPE ATCE*, San Antonio, TX, October 5-7.

Chapter 3

A RELATION-ANALYSIS-BASED APPROACH FOR ASSESSING RISKS OF PETROLEUM-CONTAMINATED SITES IN WESTERN CANADA

Xiaosheng Qin[1], Bing Chen[2]*, Guohe Huang[1] and Baiyu Zhang[3]

[1] Faculty of Engineering, University of Regina, Regina,
Saskatchewan, Canada
[2] Faculty of Engineering and Applied Science,
Memorial University of Newfoundland, St. John's, Canada
[3] Department of Civil Engineering, Dalhousie University,
1360 Barrington St., Halifax, NS, Canada

ABSTRACT

Effective reflection of and coping with uncertainties are essential for generating reliable risk assessment outcomes. In this study, an integrated risk assessment approach was proposed for assessing environmental risks associated with contamination of multi-component petroleum hydrocarbons. The approach consists of (a) predicting contaminant flow and transport through a multiphase multi-component numerical modelling system; (b) using interval-anlaysis approach to analyze effects of uncertainties associated with subsurface conditions; and (c) applying fuzzy relation analysis to quantify the general risks based on the interval-analysis results. The application of the proposed approach to a petroleum-contaminated site in western Canada indicated that system uncertainties would have significant impact on the risk assessment results. Implementation of risk assessment under more deterministic conditions generates clearer risk assessment outcomes, but leads to missing of more valuable information; conversely, risk assessment under more uncertain circumstances offers more comprehensive information, but leads to higher vagueness of risk descriptions. Application of the proposed approach in risk assessment of groundwater contamination represents a new contribution to the area of petroleum waste management under uncertainty. It is not only useful for evaluating risks of a system containing multiple factors with complicated interrelationships, but also advantageous in

---

* Corresponding author: Email:bingchen@engr.mun.ca

situations when probabilistic information is unavailable for performing a conventional stochastic risk assessment.

Keywords: petroleum contamination, risk assessment, fuzzy relation analysis, soil and groundwater.

1. INTRODUCTION

Risk analysis for petroleum contaminations are important tasks before identification of timely and cost-effective remediation actions (Chen et al., 1998, 1999). Implementation of an effective environmental risk assessment normally involves tasks of quantifying system uncertainties and comparing the obtained inexact system outputs with environmental guidelines (Woodbury and Dudicky, 1991). The uncertainties may be derived from measurement errors or inaccurate estimates of the subsurface conditions such as aquifer heterogeneity of soils or physiochemical and biological properties of the contaminants being released and transported. The uncertainties may also be caused by human-induced deviations such as biased judgment on using different environmental criteria on a specific contaminant; for example, the environmental standards for BTEX vary significantly among different authorities across countries. Therefore, effective reflection of and coping with these uncertainties are essential for generating reliable risk assessment outcomes.

Generally, most of the previous risk assessment is based on probability theory. Risks were measured through probability (relative likelihood) of possible contamination and magnitude (seriousness) of consequences from the contamination. Thus the risk levels could be expressed as a probability distribution over a number of adverse consequences. Many related studies in the field of groundwater contamination can be found in Thompson et al. (1992), Goodrich and McCord (1995), Bennett et al. (1998), Swartjes, (1999) and Lee et al. (2002). However, the probability theory often assumes that there exists a historical run for observation of events. In fact, when attempting to model behaviors of environmental processes, analysts often suffer from a lack of data or imperfect knowledge about the processes (Lein, 1992). As in many cases, the observation data are unavailable/incomplete, or expensive to be obtained for generating the probability distributions of parameters of interest, and have to be compromised by statistical interpolations.

Another alternative for risk assessment is based on fuzzy set theory, which is suitable for situations when probabilistic information is unavailable (Bardossy et al., 1991). A number of studies through fuzzy risk assessment for petroleum waste management have been reported. For example, Dahab et al. (1994) proposed a rule-based fuzzy-set approach for risk analysis of nitrate-contaminated groundwater by introducing fuzzy sets into a rule-based system for nitrate risk-regulation enforcement. Lee et al. (1994, 1995) proposed a fuzzy-set-based method to estimate human-health risks from groundwater contamination and evaluate possible regulatory actions. Huang et al. (1999) developed an interval parameter fuzzy relation analysis (IPFRA) model for environmental risk assessment of petroleum-contaminated aquifers due to leakage from USTs. More recently, Chen et al. (2003) applied fuzzy quantification and Monte Carlo simulation to conduct risk assessment, and the uncertainties in environmental quality criteria were addressed through fuzzy membership functions. Liu (2001) developed an integrated simulation-assessment modeling approach for analyzing cancer risks of groundwater contamination through linking an analytical groundwater solute transport model, an exposure model, and a fuzzy risk assessment (FRA) model based on fuzzy relation analysis into a general framework. Li et al. (2003) developed a hybrid fuzzy-stochastic risk assessment (FSRA) approach that can quantify both probabilistic and possiblistic uncertainties associated with site conditions, environmental quality guidelines and health impact criteria.

The previous studies have attempted to use fuzzy techniques to quantify risks in light of human-induced uncertainties; whereas, the uncertainties associated with predictions of contaminant fate and transport were still tackled by stochastic methods (e.g. Monte Carlo simulation). If no probabilistic information is available for a highly uncertain system, applications of probabilistic techniques would be significantly frustrated. Since it is typically easier for planners to define the upper and lower bounds of a variable than to define a probability distribution function, the interval analysis would be more flexible in addressing uncertainties derived from various sources (Chen et al., 1998, 1999; Maqsood, 2004). It is thus desired that an effective assessment approach be advanced for projecting uncertainties in evaluating risks associated with petroleum-contaminated sites. This study aims at developing such an approach through incorporating interval analysis and fuzzy risk assessment into a general framework, and applying it to a western Canadian case. This objective entails: (1) development of a multiphase multi-component numerical modeling system for simulating hydrocarbon contaminant flow and transport in porous media under uncertainty; (2) examination of contamination risks based on fuzzy relation analysis; and (3) application of the proposed approach to a petroleum-contaminated site in western Canada.

2. METHODOLOGY

Environmental risk assessment involves tasks of identifying source term of risk agent and its fate and transport through porous media, as well as estimating human, plant, or animal exposures and conversion of such exposures into environmental risk levels. The proposed risk assessment approach consists of three main components: (a) predict contaminant flow and transport through a multiphase multi-component numerical modeling system; (b) use interval analysis to analyze effects of uncertainties associated with subsurface conditions; and (c) use fuzzy relation analysis to quantify the general risks based on the interval-analysis results. The detailed procedures are described as follows.

2.1. Modeling Hydrocarbon Contaminant Transport in Subsurface

A critical step in understanding the behavior of NAPLs in the subsurface is a 3-D modeling analysis of the flow and transport of NAPLs and the fate of its crucial constituents. A complete description of the flow and transport in the subsurface must include (a) flow of the fluid phases including water, NAPL and gas, (b) mass transfer of components between these phases, and (c) and transport of components in each of the phases. The multiphase compositional simulators, therefore, are recognized as effective tools in investigating such complicated processes. The basic mass conservation equation for concerned components in the subsurface can be written as (Brown, 1993; Delshad et al., 1996):

$$\frac{\partial}{\partial t}(\phi \tilde{C}_k \rho_k) + \vec{\nabla} \cdot [\sum_{l=1}^{n_p} \rho_k (C_{kl} \vec{u}_l - \phi S_l \vec{\vec{D}}_{kl} \cdot \vec{\nabla} C_{kl})] = R_k \tag{1}$$

where k is the component index, l is the phase index, ϕ is the soil porosity, $\tilde{C}_k$ is the overall concentration of component k (volume fraction), ρ_k is the density of component k [ML^{-3}], $\vec{\nabla}$ is the differential operator (or divergence), n_p is the number of phases, C_{kl} is the concentration of component k in phase l (volume fraction), $\vec{u}_l$ is the Darcy velocity of phase l [LT^{-1}], S_l is the saturation of phase l, R_k is the total source/sink term for component k (volume of component k per unit volume of porous media per unit time), and $\vec{\vec{D}}_{kl}$ is the dispersion tensor [LT^{-1}]. The overall concentration $\tilde{C}_k$ is the volume of component k summed over all phases. The dispersion tensor $\vec{\vec{D}}_{kl}$ can be expressed as (Bear, 1979):

$$\vec{\vec{D}}_{klij} = \frac{D_{m,kl}}{\tau}\delta_{ij} + \frac{\alpha_{Tl}}{\phi S_l}|\vec{u}_l|\delta_{ij} + \frac{(\alpha_{Ll} - \alpha_{Tl})}{\phi S_l}\frac{u_{li}u_{lj}}{|\vec{u}_l|} \tag{2}$$

where τ is tortuosity (defined with a value greater than 1), $D_{m,kl}$ is molecular diffusion coefficient of component k in phase l [L^{2}T^{-1}], δ_{ij} is the Kronecker delta function, α_{Ll} and α_{Tl} are respectively longitudinal and transverse dispersivities of phase l [L], u_{li} and u_{lj} are Darcy velocities of phase l in directions i and j, respectively [LT^{-1}], and $|\vec{u}_l|$ is magnitude of the vector flux for phase l [LT^{-1}].

The phase flux is calculated from the multiphase form of Darcy's law (Brown, 1993):

$$\vec{u}_l = -\frac{k_{rl}\vec{\vec{K}}}{\mu_l} \cdot (\vec{\nabla}P_l - \rho_l g \vec{\nabla}z) \tag{3}$$

where k_{rl} is the relative permeability of porous medium to phase l [L^{2}L^{-2}], $\vec{\vec{K}}$ is the intrinsic permeability tensor [L^2], μ_l is the viscosity of phase l [ML^{-1}T^{-1}], ρ_l is the density of phase l [ML^{-3}], g is the acceleration of gravity [LT^{-2}], z is vertical distance which is defined as positive downward [L] and P_l is the pressure of phase l [ML^{-1}T^{-2}].

Compositional multiphase models require multiple constitutive relations to close the system of equations, and the typical constitutive relations include pressure-saturation-permeability (p-S-k) relations. The p-S-k relation in the saturated zone can be found in Delshad et al. (1996) and Lenhard and Parker (1987). The above equations can be solved numerically through the block-centered finite difference model (Bear, 1979; Lenhard and Parker, 1987; Brown, 1993; Delshad et al., 1996).

2.2. Interval Analysis for Characterizing Uncertainties

As uncertainties associated with the flow and transport models are often unavoidable due to a variety of reasons; this would lead to difficulties in quantifying the potential impact levels of pollutants in further risk assessment (Dou et al., 1995). In this study, such uncertainties are handled as interval numbers, which will be used as inputs to the contaminant flow and transport models. The outputs from interval analysis will then be used as inputs for further risk quantifications. The interval analysis is performed by using Monte Carlo simulation, with input variables being assumed in uniform distributions. Monte Carlo simulation is effective in tackling uncertainties that can be described by probability distribution functions (PDFs). It utilizes repeated executions of numerical models to simulate stochastic processes of groundwater flow and contaminant transport (Hu and Huang, 2002). Each execution of the model produces a sample output. The output samples can then be examined statistically and distributions can be determined. The primary components of a Monte Carlo simulation include probability distribution functions, random number generator, sampling rule, scoring, error estimation, variance reduction techniques, and parallelization and vectorization (Maqsood et al., 2003; Maqsood, 2004). Monte Carlo techniques have a number of advantages, such as (1) the ability to handle uncertainty and variability associated with model coefficient, (2) it can potentially be applied in deterministic modeling structure, and (3) there is a great deal of flexibility with respect to the types of probability distributions that can be used to characterize model inputs. The final outputs are expressed as intervals, with upper and lower bounds being obtained from Monte Carlo simulation.

2.3. Fuzzy Relation Analysis for Environment Risk Assessment

The next step is to use fuzzy relation analysis to examine the risk levels based on the interval analysis results and related environmental criteria. The detailed procedures of fuzzy relation analysis can be found in Bellman and Zadeh (1970), Zadeh (1975), Lai and Hwang (1992), and Huang et al. (1999). With different pollutants being concerned, the generated concentrations from interval analysis can be presented as follows:

$$\mathbf{C}^{\pm} = \{ c_i^{\pm}(\mu) | i = 1, 2, ..., m \}, \tag{4}$$

where m is the number of pollutants; $c_i^{\pm}(\mu)$ is an interval for denoting inexact concentration of pollutant i. The detail of the $c_i^{\pm}(\mu)$ is listed as follows (Huang et al., 1999; Freissinet et al., 1999):

$$c_i^{\pm}(\mu) = [c_i^{-}(\mu), c_i^{+}(\mu)] = \{ t \in c_i^{\pm}(\mu) \mid c_i^{-}(\mu) \le t \le c_i^{+}(\mu) \}, \tag{5}$$

where $c_i^{-}(\mu)$ and $c_i^{+}(\mu)$ are the lower and upper bounds of $c_i^{\pm}(\mu)$, respectively. When $c_i^{-}(\mu) = c_i^{+}(\mu)$, $c_i^{\pm}(\mu)$ becomes a deterministic number. Thus, the risk assessment can be initiated by first defining set **U** for pollutants and set **V** for risk levels as follows:

$$U = \{ u_i \mid \forall i \}, \tag{6}$$

$$V = \{ v_j \mid \forall j \}, \tag{7}$$

where u_i represent pollutant i, and v_j is for risk level j. Fuzzy subsets of **U** and **V** can then be determined as follows:

$$\underset{\sim}{A} = a_1/u_1 + a_2/u_2 + \ldots + a_m/u_m, \tag{8a}$$

$$\underset{\sim}{B} = b_1/v_1 + b_2/v_2 + \ldots + b_n/v_n, \tag{8b}$$

where a_i represents the membership grade of u_i (for pollutant i) versus the multifactorial space, and b_j denotes an integrated membership grade for risk level j. The a_i value can generally be regarded as a weighting coefficient for pollutant i. A fuzzy subset of $\mathbf{U} \times \mathbf{V}$, which is a binary fuzzy relation from **U** to **V**, can be characterized through the following membership function:

$$\underset{\sim}{R} : U \times V \to [0, 1]. \tag{9}$$

Thus, we have fuzzy relation matrix:

$$\underset{\sim}{R} = \{ r_{ij} \mid i = 1, 2, \ldots, m; j = 1, 2, \ldots, n \}, \tag{10}$$

where r_{ij} is the membership grade of pollutant i versus risk level j, which is a function of pollutant concentration and risk level criteria.

Since $c_i^{\pm}(\mu)$ is an interval number, a parametric approach that improves Kuzmin's methods (Kuzmin, 1981) is proposed to determine membership functions for interval terms in $\mathbf{C}^{\pm} = \{c_i^{\pm}(\mu) \mid \forall i\}$. Kuzmin's technique defines membership as a bilinear function that requires three parameters: α, β and γ, with $\alpha < \beta < \gamma$, where α and γ represent the minimum and maximum values, respectively. Thus, the membership function of input factor c from fuzzy set $\underset{\sim}{A}$, denoted as $\mu_{\underset{\sim}{A}}(c)$, is equal to unity at β, and zero at α and γ. However, since c is an interval number (i.e., $c = c^{\pm}$), we can analytically express $\mu_{\underset{\sim}{A}}(c^{\pm})$ as follows.

(i) when $\alpha \leq c^{\pm} \leq \beta$:

$$\mu_{\underset{\sim}{A}}^{+}(c^{\pm}) = (c^{+} - \alpha)/(\beta - \alpha), \tag{11a}$$

$$\mu_{\underset{\sim}{A}}^{-}(c^{\pm}) = (c^{-} - \alpha)/(\beta - \alpha); \tag{11b}$$

(ii) when $\beta < c^{\pm} \leq \gamma$:

$$\mu_{\underset{\sim}{A}}^{+}(c^{\pm}) = (\gamma - c^{-})/(\gamma - \beta), \tag{11c}$$

$$\mu_A^-(c^\pm) = (\gamma - c^+)/(\gamma - \beta); \tag{11d}$$

(iii) when $c^- \leq \beta$ and $c^+ \geq \beta$:

$$\mu_A^+(\gamma^\pm) = 1, \tag{11e}$$

$$\mu_A^-(\gamma^\pm) = \min\{(c^- - \alpha)/(\beta - \alpha), (\gamma - c^+)/(\gamma - \beta)\}. \tag{11f}$$

Thus, we have $\mu_A^\pm(c^\pm) = [\mu_A^-(c^\pm), \mu_A^+(c^\pm)]$. Equations (13) to (18) are used for evaluating the risk level of a contaminant concentration under a single criterion. For the multi-level criteria problems such as risk assessment for leaking underground storage tanks, the membership grade of fuzzy relation between given $c_i^\pm$ and risk level j can be calculated as follows.

(i) when $v_{i,j-1} \leq c_i^\pm \leq v_{ij}$:

$$r_{ij}^+ = (c_i^+ - v_{i,j-1})/(v_{ij} - v_{i,j-1}), \quad \forall\, i,j, \tag{12a}$$

$$r_{ij}^- = (c_i^- - v_{i,j-1})/(v_{ij} - v_{i,j-1}), \quad \forall\, i,j, \tag{12b}$$

(ii) when $v_{i,j} \leq c_i^\pm \leq v_{i,j+1}$:

$$r_{ij}^+ = (v_{i,j+1} - c_i^-)/(v_{i,j+1} - v_{i,j}), \quad \forall\, i,j, \tag{12c}$$

$$r_{ij}^- = (v_{i,j+1} - c_i^+)/(v_{i,j+1} - v_{i,j}), \quad \forall\, i,j, \tag{12d}$$

(iii) when $c_i^\pm \leq v_{i,j-1}$ or $c_i^\pm \geq v_{i,j+1}$:

$$r_{ij}^+ = 0, \quad \forall\, i,j; \tag{12e}$$

(iv) when $c_i^- \leq v_{ij}$ and $c_i^+ \geq v_{ij}$:

$$r_{ij}^+ = 1, \quad \forall\, i,j; \tag{12f}$$

$$r_{ij}^- = \min\{(c_i^- - v_{i,j-1})/(v_{ij} - v_{i,j-1}), (v_{i,j+1} - c_i^+)/(v_{i,j+1} - v_{ij})\}, \forall\, i,j; \tag{12g}$$

where v_{ij} is the criterion for pollutant i at risk level j. Thus, we can obtain the following interval parameter fuzzy relation matrix:

$$\underline{\mathbf{R}}^{\pm} = \{ r_{ij}^{\pm} \mid i = 1, 2, \ldots, m; j = 1, 2, \ldots, n \}. \tag{13}$$

Similarly, the weighting coefficients could be deterministic or interval numbers. Thus, we have:

$$\underline{\mathbf{A}}^{\pm} = \{ a_i^{\pm} \mid i = 1, 2, \ldots, m \}, \tag{14}$$

Thus, the integrated risk level can be determined as follows:

$$\underline{\mathbf{B}}^{\pm} = \underline{\mathbf{A}}^{\pm} \circ \underline{\mathbf{R}}^{\pm} \leftrightarrow \mu_{\underline{\mathbf{A}}^{\pm} \circ \underline{\mathbf{B}}^{\pm}}, \tag{15}$$

where ∘ can be a Σ-∗ or max-∗ composition (Kaufmann, 1975; Civanlar and Trussel, 1986). For example, if the Σ-∗ composition is used, we have:

$$b_j^{-} = \sum_{i=1}^{m} (a_i^{-} \ast r_{ij}^{-}), \tag{16a}$$

$$b_j^{+} = \sum_{i=1}^{m} (a_i^{+} \ast r_{ij}^{+}). \tag{16b}$$

Let $b_k^{\pm} = \max\{b_j^{\pm} \mid j = 1, 2, \ldots, n\}$. Based on the principle of maximum membership degree (Kaufmann, 1975), it can be determined that the system has an integrated risk level k. The other $b_j^{\pm}$ elements ($j \neq k$) are also useful for further specification of the risk characteristics.

3. CASE STUDY

3.1. Background

A petroleum-contaminated site located in southwestern Saskatchewan, Canada was investigated through the proposed approach. This site had been operated as processing plant to remove naphtha condensate from the natural gas stream prior to transporting to a regional transmission line. Throughout the site history, naphtha condensate, which was a waste liquid removed from the gas by a series of scrubbers, was disposed of in an underground storage tank (UST). Due to leakage of this UST in the past years, the site was seriously contaminated. During the past years, a number of field investigations were conducted to examine the hydrogeologic and environmental conditions at the site.

It was determined that the leaking UST had impacted a large surrounding area and posed serious risks to local communities (e.g. aquatic life). The groundwater sampling results indicated that the free products and concentrations of benzene, toluene, ethyl-benzene, xylenes (BTEX) were higher than the local aquatic groundwater quality guidelines. Appropriate remediation actions were needed to mitigate this kind of risk. In 2000, actions of dual phase vacuum extraction (DPVE) were conducted to clean up the free phase at this site. However, one of the main problems associated with the remediation effort is the deficiency in our knowledge of the processes controlling fluid flow and contaminant transport in groundwater. This situation was further complicated with a number of uncertainties associated with the on-site hydrogeological conditions. In addition, impacts of the contaminated site on the surrounding communities need to be quantitatively defined. The criteria of risk levels under different combinations of BTEX concentrations are provided in Table 1, which were obtained based on examinations of the related environmental guidelines and/or standards (SERM, 1999; USEPA, 1995). These risk categories reflect the degree of seriousness of the contamination impacts; in real-world applications, questionnaire surveys, round-table discussions and expert consultations would be helpful for identifying their proper levels. Thus, the problems under consideration include: (1) is current situation risky? (1) are there still any risks of BTEX contamination to local environment after 10, 25, 50, and 75 years? and (2) How serious are they?

Table 1. Criteria of Risk Levels under Different Concentrations for Each Pollutant (μg/L)

	Clean	Practically not-risky	Slightly-risky	Risky	Highly-risky
Benzene	10	50	100	370	550
Toluene	40	300	1000	1800	2200
Ethyl-Benzene	2.4	90	150	700	1000
Xylenes	100	300	1000	1800	3600

3.2. Model Development

The study system is defined as a three-dimensional domain, and the area is 180 × 150 m^2 with a depth of 20 m. Vertically, the simulation domain is discretized into four gridblocks corresponding to four layers; each layer is located at the middle of each vertical grid block, since the model applies a block-centered finite difference scheme. The vertical gridblock size is 5 m. Each layer is discretized into 30 × 25 grid blocks as shown in Figure 1. Each grid has dimensions of 6, 6, and 5 m in x, y, and z directions, respectively. The total number of grids in the 3-D computational grid system is 3000 (30 × 25 × 4). Layers 3 and 4 are located in the saturated zone, while layers 1 and 2 are in the unsaturated zone. The study site has complex heterogeneous hydrogeological conditions, possessing a mixture of three soil types including sandy soil, clay till and silty clay. The detailed profile of soil structures can be found in Li (2003). The modeling study focuses on the fate of benzene, toluene, ethyl-benzene, and xyleness (BTEX) in the aquifer system. Evaluation of potential impacts from the groundwater contamination will then be undertaken based on the modeling outputs.

The model was calibrated using the on-site information of 1998 to 1999. The monitoring data of 2000 were used to verify the developed model. Table 2 lists part of the estimated modeling inputs.

The other modeling inputs were obtained from a number of sources, e.g. reports of the previous site investigation and laboratory analysis. The verification figures can refer to literatures of Li (2003) and Huang et al. (2007). It was found that the simulated maximum BTEX concentrations were 0.429, 1.423, 0.987, and 1.222 mg/L, respectively; while the observed ones were 0.345, 1.3, 0.602, and 1.120 mg/L, respectively. The relative errors were about 19.0%, 8.6%, 38.9%, and 8.2%, respectively. The results demonstrated that differences between the predicted and observed BTEX concentrations were generally acceptable.

The numerical models used in this study could reasonably simulate the fate and transport of the contaminants. The verified models can then be used for investigating the effects of parameter uncertainties on BTEX concentrations and performing risk assessment in surrounding environment. With a conservative consideration, the peak BTEX concentrations 10 years later at layer 3 will be used for further risk assessment. This is based on the assumption that the impact of a contaminant at its peak concentration will pose greater threats than those at lower concentrations. Layer 3 is selected as it is the first layer of the saturated zone and more risky than other layers. The 10-year time horizon is selected as an example.

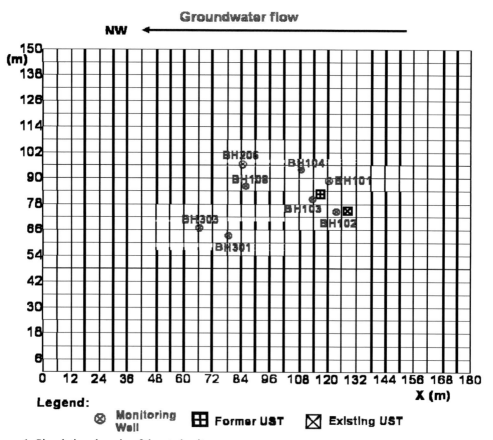

Figure 1. Simulation domain of the study site.

Table 2. Modeling input parameters

Parameter	Value	Unit
Permeability of sandy soil in x, y, and z direction	1900	MD
Permeability of clay till in x, y, and z direction	195	MD
Permeability of silty clay in x, y, and z direction	380	MD
Porosity of sandy soil	0.35	--
Porosity of till	0.30	--
Porosity of silty clay	0.53	--
NAPL/water interfacial tension	45	Dynes/cm
NAPL density	0.713	g/cm^3
Longitudinal dispersivity	5	m
Transverse dispersivity	0.5	m
Hydraulic gradient	0.006	m/m
Benzene solubility	1750	mg/l
Ethylbenzene solubility	152	mg/l
Toluene solubility	535	mg/l
Xylenes solubility	175	mg/l

3.3. Result Analysis

As indicated in Table 1, the soil structures (mixture of sandy soil, clay till, and silty clay) have significant uncertainties, leading to imprecise modeling inputs. In this study, two critical parameters, i.e. intrinsic permeability (K_{xx}) and the porosity (n), are considered uncertain and independent (Dong and Shah, 1987). They will be tackled as interval numbers. Each interval number is defined by specifying the maximum credible value (1000 milli Darcy for K_{xx} and 0.4 for n), and the lowest (200 milli Darcy for K_{xx} and 0.2 for n) possible values (1800 milli Darcy for K_{xx} and 0.6 for n).

Figure 2 shows the intervals of BTEX peak concentrations in layer 3 after 10, 25, 50 and 75 years, respectively. It is indicated that the uncertainties in input parameters of K_{xx} and n will significantly impact the predicted modeling outputs. For example, after 25 years, the toluene concentration would range from 710.3 to 1213.6 μg/L, demonstrating a signficiant variation; benzene and ethyl-benzene concentrations fluctuate in the range of 203.1 to 339.4 μg/L and 267.2 to 446.5 μg/L, respectively, presenting relatively lower but still considerable variations.

These facts indicate that the impacts of uncertainties could possibly lead to biased or even false estimation of the risk levels of contaminants. For example, the average modeling output of toluene concentration after 25 years is around 960 μg/L; this toluene level would be safe according to the USEPA health limits of toluene (1000 μg/L). However, due to the existence of system uncertainties, such toluene guideline will be easily violated.

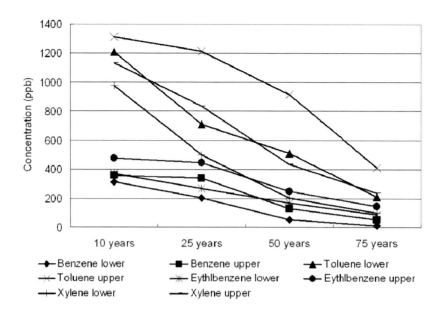

Figure 2. Predicted BTEX concentrations through interval analysis.

Results shown in Figure 2 demonstrate that the BETX concentrations are presented as interval numbers. Therefore, for the generated fuzzy interval at each time point, the parametric approach that based on Kuzmin's methods (Kuzmin, 1981) can be used to determine the interval-parameter fuzzy relation matrix. Thus, further risk assessment can be conducted through interval-parameter fuzzy relation analysis that is incorporated into the integrated risk assessment framework.

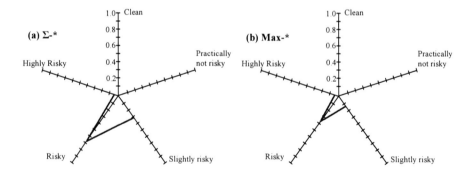

Figure 3. Environmental risk assessment of BTEX contamination at the initial conditions.

In this study, only deterministic weighting coefficiencies will be considered. Through evaluating impacts from each contaminant in contribution to the health impact, the weighting set W for BETX can be obtained as (0.40, 0.24, 0.20, 0.16) (Liu, 2001). The weighting levels can be determined based on public survey, round-table discussion or expert consultations. Figure 3 presents the results of integrated risk assessment through the proposed approach on the current BTEX distributions (initial conditions). Figures 3a and 3b are obtained by using Σ-* and Max-* compositions, respectively. It appears that the environmental conditions at the study site will be "risky" in terms of the threat against health safety in local communities. There are also probabilities that the situation becomes "highly risky" under extreme conditions, leading to significant injuries to human health. However, with the effect of natural attenuation and groundwater flow, the BTEX distributions may change considerably.

Figures 4 and 5 present the results of integrated risk assessment on the future conditions through the proposed approach, with Σ-∗ and Max-∗ compositions being used in relation analysis (Kaufmann, 1975). It is indicated that, due to natural attenuation effects, the environmental conditions at the study system will get improved. After 10 years, the general risk conditions range from "practically not risky" to "risky" level with varied system possibilities. For example, as shown in Figure 4a, the upper bound of the possibilities of "practically not risky", "slightly risky", and "risky" on study site is 0.05, 0.54 and 0.62, respectively, indicating a possibility of system risk level of "slightly risky" to "Risky". Therefore, different site remediation strategies (with different remediation efficiencies) can be evaluated with respect to the results of the proposed risk assessment approach in order to get acceptable safety levels. For example, through the proposed approach, it is indicated that a 40% remediation action for the current BTEX contaminations will improve the study system from "slightly risky" level with a possibility of 0.54 to a possibility of 0.23, and from "Risky" with a possibility of 0.62 to 0.46; while a 90% remediation action will result in an improvement of the study system to "practically not risky" with a possibility of 0.66 and "clean" with a possibility of 0.20; however, higher-efficiency remediation scheme normally based on higher system cost; thus, a suitable remediation strategy that compromises between environmental and economic objectives can be determined.

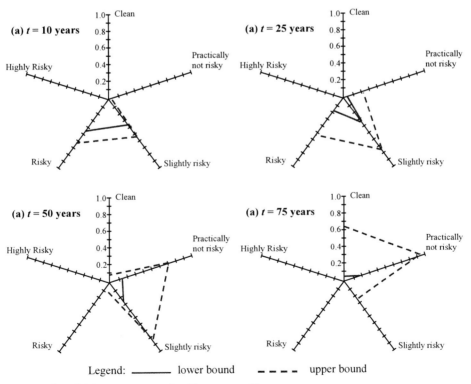

Figure 4. Results of risk assessment using Σ-∗ composition.

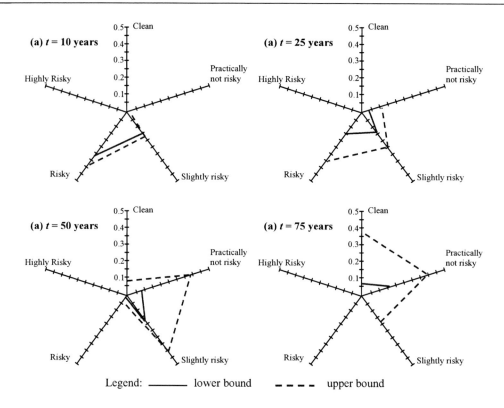

Figure 5. Results of risk assessment using Max-∗ composition.

The above results have indicated that the system uncertainties will have significant impacts on the risk assessment results. Implementation of risk assessment under more deterministic conditions generates clearer risk assessment outcomes, but leads to a missing of more valuable information.

Conversely, risk assessment under more uncertain circumstances offers more comprehensive information, but leads to higher vagueness of risk descriptions. We can hardly achieve both completeness and clearness in risk descriptions due to system uncertainties and complexities. Therefore, a tradeoff has to be made before a conclusion of risk assessment is drawn, which may be critical for guiding further remediation actions.

Generally, the proposed risk assessment approach is useful for comprehensively evaluating risks within a system containing many uncertain factors with complicated interrelationships.

For the study problem under consideration, the risk assessment approach allows engineers to have a systematic and consistent approach for assessing uncertainty impacts and environmental risks. The risk assessment outputs will thus provide bases for determining desirable site remediation actions. The advantages of using the proposed approach are: (1) it is an integrated approach that incorporates effects of different pollutants within a general framework; (2) it can effectively reflect uncertainties presented as interval numbers for a number of modeling inputs.

CONCLUSION

In this study, an integrated risk assessment approach was proposed for evaluating environmental risks associated with groundwater contamination due to leaking USTs. Results of a case study that located in western Canada indicate that reasonable solutions for risk analysis under different system conditions can be generated. Results from the study case indicated that the uncertain parameters will have large influences on modeling outputs. Risk assessment under more uncertain circumstances offers more comprehensive information, but leads to higher vagueness of risk descriptions. We can hardly achieve both completeness and clearness in risk descriptions due to system uncertainties and complexities. Therefore, a tradeoff has to be made before a conclusion of risk assessment is drawn, which may be critical for guiding further remediation actions.

The proposed methodology is advantageous in situations when any probabilistic information is unavailable for performing a conventional stochastic risk assessment. It is effective in tackling not only the problems of BTEX contaminations, but also other similar environmental problems associated with multiple contaminants. However, there are still a number of limitations that need to be addressed in further studies: (a) determination of proper weighting levels for multiple contaminants is highly subjective and should be based on extensive public survey and stakeholder discussions; (b) the risk quantification method would be less effective when the interval outputs of the contaminant concentrations are highly vague; (c) the calculation amount of interval analysis would become enormous if the number of uncertain parameters was large. The proposed approach is useful for comprehensively evaluating risks within a system containing many factors with complicated interrelationships. It can incorporate effects of different pollutants within a general framework. Also the method can effectively reflect uncertainties presented as interval numbers for a number of modeling inputs. The modeling results can provide bases for determining desirable site remediation actions. Application of the proposed in risk assessment of groundwater contamination represents a new contribution to the area of petroleum waste management under uncertainty. Further explorations based on the proposed approach would be beneficial. For example, the proposed risk assessment approach can also be used for analyzing other site contamination problems, such as leakage at gasoline production sites.

ACKNOWLEDGMENT

This article is based on the work supported by the Major State Basic Research Development Program of the MOST (2006CB403307) and the NSFC (50709010). The author would like to thank the anonymous peer reviewers for their useful comments and suggestions.

REFERENCES

Bardossy, A.; Bogardi, I.; Duckstein, L. Fuzzy set and probabilistic techniques for health-risk analysis, *Applied Mathematics and Computation*, 1991, 45, 241-268.

Bear, J. *Hydraulics of Ground Water*, McGraw-Hill, New York, 1979.

Bellman, R. E.; Zadeh, L. A. Decision making in a fuzzy environment, *Management Science*, 1970, 17, 41-64.

Bennett, D. H.; James, A. L.; McKone, T. E.; Oldenburg, C. M. On uncertainty in remediation analysis: variance propagation from subsurface transport to exposure modeling, *Reliability Engineering and System Safety*, 1998, 62, 117-129.

Brown, C. L. *Simulation of Surfactant Enhanced Remediation of Aquifers Contaminated with Dense Non-aqueous Phase Liquids*, Ph.D. Dissertation, University of Texas at Austin, TX, 1993.

Chen, Z.; Huang, G. H.; Chakma, A. Integrated environmental risk assessment for petroleum-contaminated sites - a north American case study, *Water Science and Technology*, 1998, 38, 131-138.

Chen, Z.; Huang G. H.; Chakma, A. *Methodology for subsurface modeling. In Numerical Modeling, Risk Assessment, and Site Remediation*. Technical Report, Energy and Environment Program, University of Regina: Regina, SK, Canada, 1999.

Chen, Z.; Huang, G. H.; Chakma, A. Hybrid fuzzy-stochastic modeling approach for assessing environmental risks at contaminated groundwater systems, *Journal of Environmental Engineering*, 2003, 129, 79-88.

Civanlar, M. R.; Trussel, H. J. Constructing membership functions using statistical data, *Fuzzy Sets and Systems*, 1986, 18, 1-13.

Dahab, M. F.; Lee, Y. W.; Bogardi, I. Rule-based fuzzy-set approach to risk analysis of nitrate-contaminated groundwater, *Water Science and Technology*, 1994, 30, 45-52.

Delshad, M.; Pope, G. A.; Sepehrnoori, K. A compositional simulator for modeling surfactant enhanced aquifer remediation, 1. formulation, *Journal of Contaminant Hydrology*, 1996, 23, 303-327.

Dong, W.; Shah, H. C. Vertex method for computing functions of fuzzy variables, *Fuzzy Sets and Systems*, 1987, 24, 65-78.

Dou, C.; Woldt, W.; Bogardi, I.; Dahab, I. Steady state groundwater flow simulation with imprecise parameters, *Water Resource Res.* 1995, 31, 2709-2719.

Environment Canada. *Over Seven Million Canadians Depend on Groundwater*. Rep. EC-200016, Environment Canada, Burlington, Ont., Canada, 1979.

Freissinet, C.; Vauclin, M.; Erlich, M. Comparison of first-order analysis and fuzzy set approach for the evaluation of imprecision in a pesticide groundwater pollution screening model, *Journal of Contaminant Hydrology*, 1999, 37, 21-34.

Goodrich, M. T.; McCord, J. T. Quantification of uncertainty in exposure assessments at hazardous waste sites, *Ground Water*, 1995, 33, 727-732.

Hu, B. X.; Huang, H. Stochastic analysis of reactive solute transport in heterogeneous, fractured porous media: a dual-permeability approach, *Transport in Porous Media*, 2002, 48, 1-39.

Huang, G. H.; Chen, Z.; Tontiwachwuthikul, P.; Chakma, A. Environmental risk assessment for underground storage tanks through an interval parameter fuzzy relation analysis approach, *Energy Sources*, 1999, 21, 75-96.

Huang, Y. F.; Huang, G. H.; Chakma, A.; Maqsood, I.; Chen, B. Remediation of petroleum-contaminated sites through simulation of a DPVE-aided cleanup process: part 1. Model development. *Energy Sources*, 2007, 29, 347-365.

Lai, Y. J.; Hwang, C. L. *Fuzzy Mathematical Programming Methods and Applications*, Springer-Verlag, Berlin, 1992.

Lee, L. J. H.; Chan, C. C.; Chung, C. W.; Ma, Y. C.; Wang, G. S.; Wang, J. D. Health risk assessment on residents exposed to chlorinated hydrocarbons contaminated in groundwater of a hazardous waste site, *Journal of Toxicology and Environmental Health*, Part A, 2002, 65, 219-235.

Lee, Y. W.; Dahab, M. F.; Borgardi, I. Fuzzy decision making in ground water nitrate risk management, *Water Resources Research*, 1994, 30, 135-148.

Lee, Y. W.; Dahab, M. F.; Bogardi, I. Nitrate-risk assessment using fuzzy-set approach, *Journal of Environmental Engineering*, 1995, 121, 245-256.

Lein, J. K. Expressing environmental risk using fuzzy variables: a preliminary examination, *Environmental Professional*, 1992, 14, 257-267.

Lenhard, R. J.; Parker, J. C. Measurement and prediction of saturation-pressure relationships in three-phase porous media systems, *Journal of Contaminant Hydrology*, 1987, 1, 407-424.

Li, J. B. *Development of an Inexact Environmental Modeling System for the Management of Petroleum-Contaminated Sites*, Ph.D. Dissertation, University of Regina, Regina, Saskatchewan, Canada, 2003.

Liu, L. Development of environmental modeling methodologies for the management of regional and industrial pollution control systems, Ph.D. Dissertation, University of Regina, Regina, Saskatchewan, Canada, 2001.

Kaufmann, A. *Introduction to the theory of fuzzy subsets*. New York: Wiley, 1975.

Kuzmin, V. B. A parametric approach to description of linguistic values of variables and hedges, *Fuzzy Sets and systems*, 1981, 6, 27-41.

Maqsood, I. *Development of Simulation- and Optimization-Based Decision Support Methodologies for Environmental Systems Management*, Ph.D. Dissertation, University of Regina, Regina, Saskatchewan, Canada, 2004.

Maqsood, I.; Li, J. B.; Huang, G. H. Inexact multiphase modeling system for the management of uncertainty in subsurface contamination, *ASCE Practice Periodical of Hazardous, Toxic, and Radioactive Waste Management*, 2003, 7, 86-94.

SERM (Saskatchewan Environment and Resource Management). *Risk Based Corrective Actions for Petroleum Contaminated Sites*, Province of Saskatchewan, Regina, Saskatchewan, Canada, 1999.

Swartjes, F. A. Risk-based assessment of soil and groundwater quality in the Netherlands: standards and remediation urgency, *Risk Analysis*, 1999, 19, 1235-1249.

Thompson, K. M.; Burmaster, D. E.; Crouch E. A. C. Monte Carlo techniques for quantitative uncertainty analysis in public health risk assessments. *Risk Analysis*, 1992, 12, 53-63.

USEPA (U. S. Environmental Protection Agency). *How to Evaluate Alternative Cleanup Technologies for Underground Storage Tank Sites: a Guide for Corrective Action Plan Reviewers*, EPA 510-B-95-007, Washington, DC., 1995.

Woodbury, A. D.; Dudicky, E. A. The geostatisitcal characteristics of the borden aquifer, *Water Resources Research*, 1991, 27, 533-546.

Zadeh, L. A. The concept of a linguistic variable and its application to approximate reasoning, *Information Sciences*, 1975, 8, 199-249.

In: New Developments in Sustainable Petroleum Engineering ISBN: 978-1-61324-159-2
Editor: Rafiq Islam © 2012 Nova Science Publishers, Inc.

Chapter 4

A TWO-STEP MODELING APPROACH FOR THE RECOVERY OF FREE PHASE LNAPL IN PETROLEUM-CONTAMINATED GROUNDWATER SYSTEMS

Z. Chen[*] and J. Yuan

Department of Building, Civil, and Environmental Engineering,
Concordia University, Montreal, Quebec, Canada

ABSTRACT

This study presents a two-step modeling method to the recovery of leaked petroleum product in groundwater system. Specifically, an oil volume estimation method is developed to calculate the total volume of LNAPL residing in both saturated and unsaturated zones under concern. With the information of LNAPL distribution in the groundwater system, porous media multiphase simulation technique is then used to examine the remediation alternative of vacuum-enhanced multiphase extraction (i.e., VER). VER modeling results can be also verified by the oil estimation analysis through comparing recovered and estimated LNAPL volumes. At the study site, the soil and groundwater are contaminated by leaked condensate from perforated underground storage tanks. The VER system designed for the separation and recovery of condensate from subsurface consists of one extraction well. Good results have been obtained with the recovered amount of LNAPL close to the estimated volume. It indicates the developed method is effective in simulating the recovery process of LNAPL from the contaminated site, and thus provides optimal parameters for the remediation program.

Keywords: LNAPL, petroleum, groundwater, modeling, multiphase, vacuum-enhanced oil recovery.

---

[*] Correspondance: Dr. Zhi Chen, E-mail: zhichen@alcor.concordia.ca

1. INTRODUCTION

Groundwater contamination due to light non-aqueous phase liquid (LNAPL) has been a serious environmental problem in the past decades. The fate of LNAPL in the subsurface has been the subject of intensive studies for many years. It is currently known that LNAPL' migration could be influenced by capillary, gravity, and buoyancy forces. When they are released into the subsurface and transported through a porous formation, a portion of LNAPL is retained within the pores and trapped in unsaturated and saturated zones as discrete ganglia due to capillary forces, while others may form a layer of LNAPL on the water table [1]. LNAPL in the subsurface are difficult to remediate because they can produce underground contamination in all the four phases: (1) free phase residing at the water table; (2) dissolved phase in groundwater and pore water; (3) residual phase in soil within the vadose zone, saturated zone, and capillary fringe; and (4) vapor phase volatilized and present as soil gas within the vadose zone. Among them, the dissolved phase could pose an immediate potential threat to human health through drinking water supply. In the meantime, since LNAPL are typically immiscible fluids with low aqueous solubility, the free phase LNAPL and entrapped residuals may serve as long-term sources of groundwater contamination, leading to a significant impediment to aquifer restoration [2].

This situation has stimulated the development of innovative technologies for the separation and removal of LNAPL from petroleum-contaminated sites. Among them, vacuum-enhanced oil recovery (VER) has been widely used as a promising technology for recovering LNAPL from subsurface. In a conventional LNAPL recovery system, the water table is drawn down, creating a cone of depression for collection of free LNAPL. A consequence of this strategy is that oil is smeared downward into potentially oil-free soils, increasing the volume of contaminated soil. This results in longer remediation time and the need to further treat residual contamination. The proposed VER method allows LNAPL to be lifted off the water table and released from the capillary fringe. Gradient in the oil potentials can be increased with minimum fluctuations in the groundwater tables. This minimizes smearing while using induced airflow to biovent the vadose zone of residual oil. The remediation can then be carried out faster and completely.

Considering the design of an efficient VER system for petroleum-contaminated site, the first important step is to acquire information related to the extent, distribution, and volume of LNAPL present in the residual amount in the unsaturated and saturated zone, and in the free phase amount accumulated above the saturated zone. Though site investigation may provide important information on current site conditions, an accurate method is required for the estimation of volume of free phase and residual LNAPL in the subsurface area. The following documentation has served as the rationale for this requirement: (1) in porous media, the distribution of LNAPL is primarily related to the oil, water, and air pressures, and the pore-size distribution of the medium. For the majority of soil and aquifer, they do not have a uniform pore-size distribution, so that the concept of a discrete LNAPL-water interface is not generally correct [3] and (2).

It has been reported that actual hydrocarbon thickness referring to the volume per unit area (LNAPL specific volume) is less than the measured NAPL thickness in a monitoring well [4,5]. More recently, Mould et al. [6] carried out an order-of-magnitude estimate of the volume of oil potentially leaking from aboveground storage tanks (ASTs), based on a relationship between AST age and the probability of failure due to corrosion. Reed et al. [7] developed a computer-based methodology for estimation of offshore pipeline oil spill volumes. Vogler et al. [8] proposed a new method based on experiments to interpret fluid levels in groundwater monitoring wells in dynamic aquifers and to estimate spilled oil volume.

Mathematical modeling is often used in predicting field performance of various remediation scenarios, based on information of site conditions as well as the extent, distribution, and volume of residual and free phase LNAPL in the system. Previously, even though the VER technology was applied to several studies on remediation of LNAPL-contaminated sites, less efforts have been devoted to the modeling for LNAPL' recovery. Many of them suffered from over-simplified assumptions under multiphase flow modeling process [9,10,11]. For example, Baker and Bierschenk [10] gave a brief description of one-, two-, and three-phase flow during free product recovery through a vacuum-enhanced method.

They also introduced the relevant concept and field test results of vacuum-enhanced recovery of water and NAPL. Hoeppel et al. [12] applied VER technology to the remediation of Naval middle distillate fuel sites. Crone et al. [13] studied the multiphase flow in homogeneous porous media with phase change for soil remediation. Given the complexities associated with multiphase flow system, few studies contain an effective measure to verify modeling results.

As an extension to previous studies, a two-step modeling approach is proposed in this study which involves: (1) an integrated oil volume estimation program to compute free phase LNAPL volume and trapped oil volumes in saturated and unsaturated zones due to water table fluctuations in the petroleum-contaminated site, and (2) a multiphase flow model to predict field performance of various remediation scenarios involving vacuum-enhanced oil recovery remediation technology based on the estimated total oil volume. The two-step modeling approach will be applied to a case study in the western Canada and site heterogeneous characteristics will be considered and incorporated within the simulation process.

2. METHODOLOGY

2.1. Oil Volume Estimation

The oil volume to be estimated in contaminated subsurface comes from three sources: the free phase oil, residual oil in unsaturated zone, and residual oil in saturated zone. The methods utilized to calculate them are described in follows, respectively.

Free Phase Oil in the Subsurface

The three-phase (air, water, LNAPL) pressure distribution under physical equilibrium at vertical direction is applicable in the monitoring well and adjacent media [3,14]. Under such conditions, the van Genuchten [15] model can be extended to predict free phase LNAPL saturation (S_{of}) based on the measured fluid elevations. The modified van Genuchten model is given by Parker et al. [16]:

$$S_w = (1 - S_m) [1 + (\alpha\, \beta_{ow}\, h_{ow})^n]^{-m} + S_m \qquad (1a)$$

$$S_t = (1 - S_m)[1 + (\alpha \beta_{ao} h_{ao})^n]^{-m} + S_m \qquad (1b)$$

where S_w is water saturation; S_t is total liquid saturation (oil + water); S_m is water saturation at soil field capacity; α, n, and m are van Genuchten parameters, and β_{ow} and β_{ao} are fluid scaling factors; the capillary heads for oil-water (h_{ow}) and air-oil (h_{ao}, [L]) can be defined as $h_{ao} = h_a - h_o$ and $h_{ow} = h_o - h_w$ [L] (Lenhard and parker, 1990); h_w and h_o are calculated by $h_w = P_w/g\rho_w$ and $h_o = P_o/g\rho_w$; and P_w and P_o are phase pressures [FL^{-2}];

Through the manipulation of above equations, S_{of} is represented as a function of elevation. Parker and Lenhard [3] proposed a vertically integrated model to calculate the free oil specific volume (V_{of}) as following integration:

$$V_{of} = \phi \int_{Z_{ow}}^{Z_u} S_{of} dz \qquad (2)$$

where $S_{of} = S_t - S_w$, ϕ is the soil total porosity; V_{of} denotes the volume of free oil per unit area of porous media [L]; Z_{ow} is the oil-water table elevation [L]; and Z_u [L] is the maximum elevation given by:

$$Z_u = Z_{ow} + (\rho_{ro} \beta_{ao} H_o)/(\rho_{ro}\beta_{ao} - (1 - \rho_{ro})\beta_{ao} \qquad (3)$$

Residual Oil in Unsaturated Zone

When the air-oil table P_{ao} falls, oil is trapped in the unsaturated regions. The maximum specific volume of oil trapped in the unsaturated region is:

$$V_{um} = S_{og} \phi \Delta P_{ao} \qquad (4)$$

where $P_{ao} = Z_{ao} - h_a$ [L] and Z_{ao} is the air-oil table elevation [L]; S_{og} is the maximum allowed unsaturated zone residual oil saturation and ΔP_{ao} is the change in the P_{ao} elevation [L]. S_{og} is expressed as:

$$S_{og} = F_g S_m (1 - S_m) \qquad (5)$$

where F_g is a function of the volume of free phase oil, porosity, and oil thickness, it ranges from 0.2 to 0.5.

Residual Oil in Saturated Zone

When the oil-water table, Z_{ow}, rises, oil is trapped in the saturated region. The maximum volume of oil trapped in the saturated region, V_{sm}, is expressed as:

$$V_{sm} = S_{or} \phi \Delta Z_{ow} \qquad (6)$$

where ΔZ_{ow} is the change in the Z_{ow} elevation [L]. S_{or} is the maximum allowed saturated region residual saturation and is written as:

$$S_{or} = F_r (1 - S_m) \qquad (7)$$

where F_r is a function of the free phase oil volume, porosity, and oil thickness, which ranges from 0.2 to 0.5.

Total Oil Volume in the Subsurface

In order to calculate total oil volume, a two-dimensional krigging algorithm is employed to interpolate H_o values from the monitoring well network into a computational mesh such that the total volume is obtained:

$$V = V_{of} + V_u + V_s = A \left[\sum_{i=1}^{N} V_{ofi} + \sum_{i=1}^{N} V_{ui} + \sum_{i=1}^{N} V_{si} \right] \tag{8}$$

where A is the representative area of a node in the computational mesh [L^2]; N is the number of nodes; V_{ofi} is the free oil specific volume at node i [L]; V_{si} is the specific volume of residual oil in saturated zone at node i [L]; and V_{ui} is the specific volume of residual oil in unsaturated zone at node i [L]; V is the total oil volume [L^3].

2.2. Multiphase Flow in the Porous Medium

Based on Darcy's equation and assuming negligible compressibility of water and oil phases as well as soil matrix, the mass conservation equations for incompressible water (w), LNAPL (o) and compressible gas (a) is given as [17,18,19]:

$$\phi \, \partial S_w / \partial t = -\partial q_{wi} / \partial x_i + R_w' \tag{9a}$$

$$\phi \, \partial S_o / \partial t = -\partial q_{oi} / \partial x_i + R_o' \tag{9b}$$

$$\phi \, \partial \rho_a S_a / \partial t = -\partial \rho_a q_{ai} / \partial x_i + R_a' \tag{9c}$$

where ϕ is the porosity [$L^3 L^{-3}$], S_p is p-phase saturation [$L^3 L^{-3}$] (p = w, o, and a for water, oil, and gas, respectively), t is time [T], R_w', R_a', and R_o' are p-phase volumetric source-sink terms [$L^3 L^{-3} T^{-1}$], x_i (or x_j) are Cartesian coordinates (i, j = 1, 2, 3), and q_{pi} are Darcy's velocities within p-phase in direction i.

In the process of model development, it is necessary to define the flow domain of interest as being bounded at the top by the soil-atmosphere boundary and at the bottom by an effective lower bound of the unconfined aquifer. Under the assumption of vertical equilibrium, integration of mass conservation Equations (9a), (9b) and (9c) over the vertical domain yields:

$$\partial V_w / \partial t = \partial (T_{wij} \, \partial z_{aw} / \partial x_j) / \partial x_i + R_w \tag{10a}$$

$$\partial V_a / \partial t = \partial (T_{aij} \, \partial z_{aa} / \partial x_j) / \partial x_i + R_a \tag{10b}$$

$$\partial V_o / \partial t = \partial (T_{oij} \partial z_{ao} / \partial x_j) / \partial x_i + R_o \tag{10c}$$

where V_w, V_a, and V_o are the specific volume of water, air, and LNAPL [L], respectively; T_{wij}, T_{aij}, and T_{oij} are water, air and oil transmissivity tensors [L^2T^{-1}], respectively; z is elevation of phase interface [L]. With the introduction of fluid capacities C_{pq} ($C_{pq} = \partial V_p/\partial z_{aq}$, q = a, o, w), Equations (11a), (11b) and (11c) can be written as the following standard form:

$$C_{ww} \partial z_{aw}/\partial t + C_{wo}\partial z_{ao}/\partial t + C_{wa}\partial z_{aa}/\partial t = \partial(T_{wij} \partial z_{aw}/\partial x_j)/\partial x_i + R_w \qquad (11a)$$

$$C_{ow} \partial z_{aw}/\partial t + C_{oo}\partial z_{ao}/\partial t + C_{wa}\partial z_{aa}/\partial t = \partial(T_{oij} \partial z_{ao}/\partial x_j)/\partial x_i + R_o \qquad (11b)$$

$$C_{ao} \partial z_{ao}/\partial t + C_{aa}\partial z_{aa}/\partial t = \partial(T_{aij} \partial z_{aa}/\partial x_j)/\partial x_i + R_a \qquad (11c)$$

The above equations are solved based on the constitutive relations among pressure, hydraulic conductivity, and saturation for three phases [16]. In this study, the outputs from the oil estimation analysis are similar to the input requirements of the multiphase model. Therefore, the multiphase extraction model can either be used alone or run together with the oil volume estimation process.

3. CASE STUDY

3.1. Overview of the Study Site

Figure 1 provides an overview of the study site (Latitude: 51.607223 °, Longitude: -109.742166 °), which is located in western Canada. The site and facilities were served and operated as a natural gas processing plant from the middle of 1960s until 1990. The plant utilized a series of scrubbers to remove condensate, recognized as a LNAPL, from the natural gas stream prior to transport in a regional transmission line. Throughout the operating history of the site, the waste condensate had been disposed of in three perforated underground storage tanks (USTs). From 1992 to 1998, a number of site investigation and monitoring works were conducted.

Useful information in relation to site conditions and contamination situations was obtained in preparation of data and provision of important bases for further simulating separation and recovery of condensate from the study site.

Groundwater was encountered between 5 and 11 m below surface. Around UST2 and UST3 area, the groundwater table was encountered to be 7 to 11 m below surface, while that at the north of the UST3 around wells BH-9 and BH-10 was 5 to 6 m deep. It is indicated that the general groundwater flow direction is towards south, with the gradient of water table being slightly from northeast to southwest. The only exception is at the site's immediate south, where higher gradients can be found at some spots. The groundwater table is predominately located within the clay-till soils.

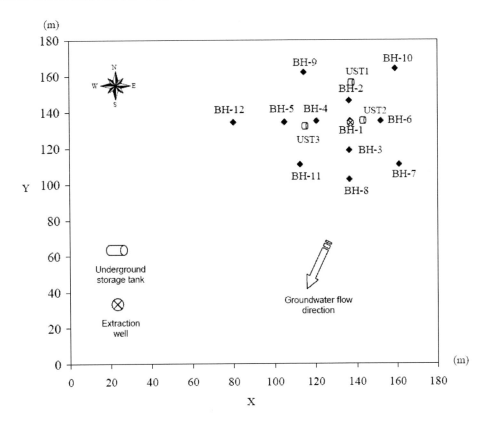

Figure 1. Overview of the study domain.

3.2. Data Investigation

A large amount of data is needed for the modeling study. The site hydrologic parameters include hydrologic properties and porous medium properties. The hydrologic properties include water dynamic viscosity, water density, water surface tension, average recharge rate, water saturation, air entry head, residential water saturation, van Genuchten's alpha, van Genuchten's n, etc. The porous medium properties include hydraulic conductivity, porosity, bulk density, surface ground elevation, groundwater elevation, groundwater gradient and direction. The hydrocarbon source parameters include hydrocarbon phase properties, dissolved constituent properties, and hydrocarbon release information. The hydrocarbon phase properties include LNAPL density, LNAPL dynamic viscosity, hydrocarbon solubility, aquifer residual NAPL saturation, vadose zone residual LNAPL saturation, soil/water partition coefficients, and LNAPL surface tension.

3.3. Simulation and Results

The two-step modeling approach was applied to the study domain for designing a vacuum-enhanced condensate recovery system. The VER system consist of one extraction well was scheduled to operate for one year. The dual fluid recovery well (as shown in Figure

1) was installed at (136.84, 134.96), which is the location of monitoring well BH-1. At the extraction well location, a gas recovery well was installed to enhance recovery and a vacuum of –3.0 m (equivalent head of water) was applied.

An areal domain of 175 × 180 m was simulated in this study, which was defined based on site investigation of plume extension. During site investigations, air-oil and oil-water fluid tables (Z_{ao} and Z_{ow}) elevations were recorded in 12 monitoring wells. An average representative value for these fluid tables is given in Table 1 as well as the x- and y-coordinates of each monitoring well. The properties of condensate include (i) ratio of oil to water dynamic viscosity (=2), (ii) ratio of oil to water phase density (= 0.73), (iii) air-oil phase scaling parameter (=3), and (iv) Oil-water phase scaling parameter (=1.5). During the simulation, the east and west boundaries were treated as no flow boundaries for water. All sides were treated as no flow for the oil phase. Along four sides of the domain, gas pressure was atmospheric.

Table 1. Fluid table elevation (m) in the monitoring wells

Monitoring wells	X (m)	Y (m)	Z_{ao} (m)	Z_{ow} (m)
BH-1	136.84	134.96	92.31	90.13
BH-2	136.46	146.24	90.66	90.55
BH-3	136.47	118.80	90.17	89.97
BH-4	120.68	134.96	90.20	90.11
BH-5	104.89	134.34	90.18	89.93
BH-6	151.88	134.96	90.37	90.37
BH-7	160.96	110.90	90.01	90.01
BH-8	136.47	102.63	89.93	89.93
BH-9	114.41	162.03	93.75	93.75
BH-10	159.02	163.91	94.49	94.49
BH-11	112.41	110.90	90.06	90.05
BH-12	80.16	134.34	90.84	90.84

Based on the provided parameters of the study domain, the simulation of vacuum-enhanced recovery of leaked condensate from the contaminated study site is carried out through the use of the proposed models:

(1) With the measured soil vapor concentrations and free phase condensate thicknesses in 12 monitoring wells, the initial leaked condensate volume from three perforated USTs at the contaminated site is calculated with a value of 12.72 m^3 the oil volume estimation method. Meanwhile, the initial distribution of specific oil values is taken as input to the multiphase flow model.

(2) The simulation outputs from the modeling of vacuum-enhanced multiphase extraction process include final air-water elevation (m) distribution and final air-oil elevation (m) distribution, initial and final specific free phase oil volume (m) distribution (Figure 2), final residual specific volume (m) distribution in unsaturated zone and saturated zone (Figure 3 and 4), as well as final total volume of recovered condensate. The final total volume of recovered condensate is 10.34 m^3.

Comparing the recovered condensate with the estimated total volume, it is indicated that almost all of the leaked condensate can be recovered after the one-year extracting operation. The recovered amount is relatively less than the estimated total amount partly because that the traditional pumping process is ineffective in extracting the residual phase LNAPL from the porous media. For the study site, only one extraction well is sufficient for the recovery and separation of free products onsite with economical and environmental efficiencies. Information of the final residual specific volume (m) distribution in unsaturated zone and saturated zone for the study site will help to determine if any further remediation actions such as bioremediation are needed after oil recovery.

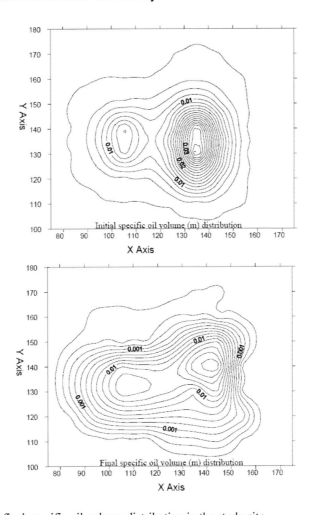

Figure 2. Initial and final specific oil volume distribution in the study site.

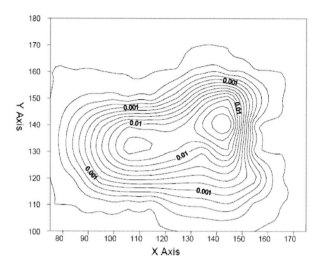

Figure 3. Final residual specific oil volume (m) distribution in saturated zone.

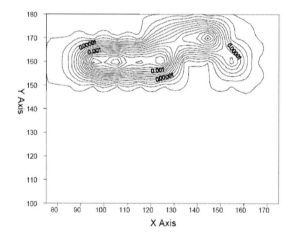

Figure 4. Final residual specific oil volume (m) distribution in unsaturated zone.

CONCLUSION

In this study, a two-step modeling approach has been developed for simulating the separation and recovery of LNAPL from soil and groundwater through the VER process at petroleum-contaminated sites. The method invokes an assumption of near-equilibrium conditions in the vertical direction. This could reduce the nonlinearity in the constitutive model and transforms a 3D problem into a 2D areal problem, thereby, drastically reducing computational time for the simulation.

Estimation of volume of LNAPL and identification of LNAPL volume distribution in the subsurface provides better understanding of typical petroleum-contaminated groundwater system. It also offers important input for the effective modeling of VER process under different settings. With the information of LNAPL distribution in both saturated and unsaturated zones in possible free and residual phases, porous media multiphase simulation technique is used to design and examine the remediation alternative of vacuum-enhanced multiphase extraction (i.e., VER). Importantly, VER modeling results can be easily verified by the oil estimation method through comparing recovered and estimated LNAPL volumes.

At the study site, the soil and groundwater are contaminated by leaked condensate from perforated underground storage tanks. The VER system designed for the separation and recovery of condensate from subsurface consists of one extraction well. The estimated LNAPL amount from the oil volume model is in agreement with the recovered amount obtained by the multiphase flow modeling. It indicates that the developed method is effective in examining the LNAPL recovery process and thus provides optimal parameters for the site remediation program.

REFERENCES

[1] Wilson, D.J. (1989). Soil clean up by in-situ surfactant flushing: mathematical modeling. *Separation Science and Technology*, 24, 863-892.

[2] Mackay, D.M., and Cherry, J.A. (1989). Groundwater contamination: pump-and-treat remediation. *Environmental Science and Technology*, 23, 630-636.

[3] Lenhard, R.J., and Parker, J.C. (1990). Estimation of free hydrocarbon volume from fluid levels in monitoring wells' *Ground Water,* 28, 57-67.

[4] Ballestero, T.P., Fiedler, F.R., and Kinner, N.E. (1994). An investigation of the relationship between actual and apparent gasoline thickness in a uniform sand aquifer. *Ground Water*, 32, 708-718.

[5] Blake, S.B., and Hall, R.A. (1984). Monitoring petroleum spills with wells: some problems and solutions. In: *Proc. Fourth Natl. Symp. On Aquifer Restoration and Ground Water Monitoring, National Ground Water Association*, Dublin, OH, 305-310.

[6] Mould, K., Watson, S., Renga, M., and Siegmann, L. (2005). Estimating the volume of oil leaking from aboveground storage tanks. In: *Proc. 2005 International Oil Spill Conference*, IOSC 2005, Miami, Florida, 2653-2657.

[7] Reed, M., Emilsen, M.H., Hetland, B., Johansen, I., Buffington, S., and Hoverstad, B. (2006). Numerical model for estimation of pipeline oil spill volumes. *Environmental Modelling and Software*, 21, 178-89.

[8] Vogler, M., Arslan, P., and Katzenbach (2001). The influence of capillarity on multiphase flow within porous media: A new model for interpreting fluid levels in groundwater monitoring wells in dynamic aquifers' *Engineering Geology*, 60, 149-158.

[9] Blake, S.B., and Gates, M.M. (1986). Vacuum-enhanced hydrocarbon recovery: a case study. In: *Proc. Petroleum Hydrocarbons and Organic Chemicals in Ground Water*, Houston, TX. November 1986. NGWA/API, Dublin, OH, 709-721.

[10] Baker, R.S., and Bierschenk, J. (1995). Vacuum-enhanced recovery of water and NAPL: concept and field test. *Journal of Soil Contamination*, 4, 57-76.

[11] Baker, R.S., and Bierschenk, J. (1996). Bioslurping LNAPL contamination. *Pollution Engineering*, 3, 38-40.

[12] Hoeppel, R.E., Kittel, J.A., and Goetz, F.E. (1995). Bioslurping technology applications at naval middle distillate fuel remediation sites. In: Hinchee, R.E., J.A. Kittel, and H.J. Reisinger (eds.) *Applied Bioremediation of Hydrocarbons*, Battelle Press, Columbus, OH, 335-346.

[13] Crone, S., Bergins, C., and Strauss, K. (2002). Multiphase flow in homogeneous porous media with phase change. *Part I: Numerical modeling. Transport in Porous Media*, 49, 291-312.

[14] Farr, A.M., Houghtalen, R.J., and McWhorter, D.B. (1990). Volume estimation of light nonaqueous phase liquids in porous media. *Ground Water*, 28, 48-56.

[15] Van Genuchten, M.Th. (1980). A closed-form solution for predicting the hydraulic conductivity of unsaturated soil. *Soil Science Society of America Journal*, 44, 892-898.

[16] Parker, J.C., Lenhard, R.J., and Kuppusamy, T. (1987). A parametric model for constitutive properties governing multiphase flow in porous media. *Water Resources Research,* 23, 892-898.

[17] Hassanizadeh, S.M., and Gray, W.G. (1979a). General conservation equations for multi-phase systems 1. Averaging procedure. *Advances in Water Resources*, 2, 131-144.

[18] Hassanizadeh, S.M., and Gray, W.G. (1979b). General conservation equations for multi-phase systems 2: mass, moment, energy, and entropy equations. *Advances in Water Resources*, 2, 191-203.

[19] Hassanizadeh, S.M., and Gray, W.G. (1980). General conservation equations for multi-phase systems 3. Constitutive theory for porous media flow. *Advances in Water Resources*, 3, 25-40.

Chapter 5

IMPROVED MODEL FOR PREDICTING FORMATION DAMAGE INDUCED BY OILFIELD SCALES

Fadairo A. S. Adesina[*1] and Omole Olusegun[2]
[1]Department of Petroleum Engineering, Covenant University, Ota, Nigeria
[2]Department of Petroleum Engineering, University of Ibadan, Nigeria

ABSTRACT

The process of formation damage due to oilfield scale precipitation and accumulation has been quantitatively modelled based on existing thermodynamics and deposition kinetic model. A variety of models on formation damage due to solid precipitation in porous media during water flooding have been reported in the literatures. Early models were based on chemical reaction involving dissolution/precipitation while neglecting the effect of operational and reservoir/brine parameters. This paper presents modified models for predicting permeability damage due to oilfield scale precipitation. The key operational and reservoir parameters which influence the magnitude of flow impairment by scale deposition were identified through the modification.

NOMENCLATURES

B = Formation volume factor, dimensionless
C = Salt Concentration, g/m^3
$C(I)$ = Concentration at the well bore pressure, g/m^3
$C(P)$ = Concentration at the reservoir pressure, g/m^3
F = Model parameter, sec^{-1}
h = Thickness, m
K_s = Instantaneous permeability, m^2
K_o = Initial permeability, m^2

---

[*] E-Mail: adesinafadairo@yahoo.com

K_{dep} = Deposition rate constant m³/gsec

K_{sp} = Solubility product M²

P = Pressure, Pa

ΔP_s = Additional pressure drop across the skin, Pa

∂P = Change in pressure, Pa

q = Flow rate, m³/sec

r_s = Radial distance, m

r_w = Well bore radius, m

S_s = Saturation of sulfate (Scale), dimensionless

S_{wi} = Connate water saturation, dimensionless

s = Skin factor, dimensionless

t = Production time, sec

T = Temperature, K

V = Volume of scale, m³

ϕ = Instantaneous porosity, dimensionless

ϕ_o = Initial Porosity, dimensionless

ϕ_d = Damaged fraction, dimensionless

ρ = Density, g/m³

μ = Viscosity, pa-sec

δ = Activity coefficient

λ_K = Permeability damage coefficient

λ_ϕ = Porosity damage coefficient

INTRODUCTION

Formation permeability damage due to oilfield scale precipitation and deposition in a porous media is a major problem during water flooding project; if the injected and formation water are incompatible (Crabtee et al., 1999; Oddo and Tomson, 1991; Omole and Osoba, 1984; Atkinson et al., 1991; Bedrikovetsky et al., 2003; Fadairo, 2004; Omole and Fadairo, 2007). Oilfield scale prediction and prevention requires the description and classification of mixing, precipitation, build up, formation porosity change and permeability damage scenarios as functions of operational and reservoir/ brine parameters. Mixing of incompatible waters occur in water-contacted portion of the reservoir while precipitation and deposition take place in the near well regions. Oilfield scale precipitation and accumulation cause formation damage in the near well region while the effect is less significant outside the vicinity of the well bore (Bedrikovetsky et al., 2003; Fadairo, 2004; Omole and Fadairo, 2007). This phenomenon provokes productivity decline by causing formation damage around the well bore region (Fadairo, 2004; Omole and Fadairo, 2007) and resisting flow in tubing, flow line and production facilities Bedrikovetsky et al., 2003). Several authors (Moghadasi et al., 2006a; Kalantari et al., 2006; Moghadasi et al., 2006b; Moghadasi et al., 2005; Frank et al., 1991; Chang and Civan, 1996; Civan, 1989) have developed models on permeability reduction due to migration of fine particles in porous media and indicated that permeability damage is more likely to be severe near the well bore. Among other authors, Rachon et al. (Moghadasi et al., 2006a; Kalantari, 2006) presented a relationship between initial permeability and instantaneous permeability as a porosity exponential function. Reis and Acock (Moghadasi et al., 2006a) developed a power law model that is valid for a solid mineral deposition induced permeability reduction of up to about Civian et al (Frank et al., 1991; Chang and Civan, 1996; Civan et al., 1989) 80%. Civian et al (Frank et al., 1991; Chang and Civan, 1996; Civan et al., 1989) presented a relationship between the initial permeability and instantaneous permeability as functions of altered porosity and initial porosity, by assuming power law of 3.0. Moghadasi et al (Moghadasi et al., 2005, 2006a, 2006b; Kalantari et al., 2006) modified the Civian et al. (Frank et al., 1991; Chang and Civan, 1996; Civan et al., 1989) model by introducing variable parameters such as particle concentration in the fluid, solid particles density against the depth and time of invasion. Omole and Fadairo (2007) and Fadairo, 2004) identified the key operational and reservoir/ brine parameters which influence the magnitude of scale build up around the well bore through the derivation of an analytical expression for minerals scale saturation as reported in appendix A. This study presents an improved model of Moghadasi et al (Moghadasi et al., 2005, 2006a, 2006b; Kalantari et al., 2006) and Civian et al.'s (Frank et al., 1991; Chang and Civan, 1996; Civan et al., 1989) formulation on formation permeability damage due to mass transfer of particles flowing through the porous media. Their formulations were modified to handle scale precipitation by considering the effect of deposition kinetic, operational and reservoir/brine parameters such as scale concentration in the brine, viscosity of brine, formation volume factor of the brine, solid scale density, injection rate, pressure drawdown, reservoir temperature, reservoir thickness, brine velocity against injection time and radial distance from the well bore vicinity.

METHODOLOGY

In developing the model, the following fundamental and general assumptions were considered:

- Solid precipitates are uniformly suspended in an incompressible fluid
- The porous medium is homogeneous isothermal and isotropic.
- The porous media contain a large number of pore spaces, which are interconnected by pore throat whose size are log–normally distributed.
- The interaction forces between the medium and precipitated solid minerals are negligible.

The Model

Consider the radial flow of a fluid at constant rate q, saturated with solid state particle at a location r, from the well bore. Assuming an idealized flow equation, Omole and Fadairo

(2007) and Fadairo (2004) expressed the pressure gradient due to the presence of scale in the flow path as follows:

$$\frac{dP}{dr} = \frac{qB\mu \exp(3K_{dep}.C.t)}{2\pi K_o h r_s} \quad (1)$$

where λ_k is defined as formation damage coefficient.

That is $\lambda_k = \exp(3K_{dep}.C.t)$ (2)

Instantaneous local porosity can be defined as the difference between the initial porosity and damaged fraction of the pore spaces.

That is $\phi_s = \phi_0 - \phi_d$ (3)

Damaged fraction of the pore spaces can be defined as the ratio of the volume of scale deposited to bulk volume of the porous media or the fraction of minerals scale that occupied the total volume of porous media.

That is $\phi_d = \dfrac{\text{volume of minerals scale deposited}}{\text{bulk volume of the porous media}}$ (4)

Substituting equation (4) in equation (3), we have

$$\phi_s = \phi_o - \frac{\text{volume of minerals scale deposited}}{\text{bulk volume of the porous media}} \quad (5)$$

The volume of scale ∂V which drops out and gets deposited in the volume element over the time interval, ∂t, is given as follows (Since ∂V = flow rate x time interval): (Robert, 1997; Civan, 2001)

$$dV = q.\left[\frac{dC}{\rho dP}\right]_T .dP.dt \quad (6)$$

where $\left(\dfrac{dC}{dP}\right)_T$ is defined as the change in saturated solid scale content per unit change in pressure at constant temperature.

Hence, the change in porosity due to scale deposition over time interval is given as:

$$d\phi_d = \frac{q\left(\frac{dC}{dP}\right)_T .dP .dt}{2\pi r_s drh \rho} \tag{7}$$

Substituting equation (1) into equation (7) and integrating the equation, we have

$$\phi_d = \frac{q^2\left(\frac{dC}{dP}\right)_T .B.\mu.t.\lambda_k}{4\pi^2 r_s^2 h^2 K_o \rho} \tag{8}$$

Substituting equation (8) into equation (3), we have

$$\phi_s = \phi_o - \frac{q^2\left(\frac{dC}{dP}\right)_T .B.\mu.t.\lambda_k}{4\pi^2 r_s^2 h^2 K_o \rho} \tag{9}$$

Dividing both sides of equation (9) by ϕ_o, we have:

$$\frac{\phi_s}{\phi_o} = 1 - \frac{q^2\left(\frac{dC}{dP}\right)_T .B.\mu.t.\lambda_k}{4\pi^2 r_s^2 h^2 K_o \rho \phi_o} \tag{10}$$

Defined F as the model parameter, given as

$$F = \frac{q^2 .B_w .\mu_w .\lambda_k}{4\pi^2 .r_s^2 .h^2 .\phi_o K_o \rho} \tag{11}$$

Inserting equation (11) into equation (10), we have:

$$\frac{\phi_s}{\phi_o} = \left(1 - F.\left(\frac{dC}{dP}\right)_T .t\right) \tag{12}$$

where $\frac{dC}{dP}$ is determined using thermodynamic model of each salt involved, reported by Atkinson et al. (1991). The detail is expressed by Omole and Fadairo (2007). Considering the relationship between the initial permeability and instantaneous permeability as a function of altered porosity, and the initial porosity defined by Civian et al. (Frank et al., 1991; Chang and Civan, 1996; Civan et al., 1989) as:

$$\frac{K_s}{K_o} = \left(\frac{\phi}{\phi_o}\right)^3 \tag{13}$$

Instantaneous permeability can be expressed as

$$K_s = K_o\left[1 - F\left(\frac{dC}{dP}\right)_T \cdot t\right]^{3.0} \tag{14}$$

Equation (14) expresses the effect of scale build up on permeability variation at different operational parameters and reservoir/brine parameters such as scale concentration in the brine, viscosity of brine, formation volume factor of the brine, solid scale density, injection rate, pressure drawdown, reservoir temperature, reservoir thickness, connate water saturation against injection time and radial distance from the well bore vicinity.

Model Validation

Window based software was developed to calculate the permeability and porosity induced by oilfield scale during water flooding as a function of operational and reservoir / brine parameters such as scale concentration in the brine, viscosity of brine, formation volume factor of the brine, solid scale density, injection rate, pressure drawdown, reservoir temperature, reservoir thickness, brine velocity against injection time and radial distance from the well bore vicinity. The past investigators and their models are listed in table 1 while the modified models are listed in table 2 for model validation. The data of Haarberg et al. (1992) on scale formation shown in table 3 and brine/reservoir properties (Robert, 1997) listed in table 4 were used as input into the model.

Table 1. Amount of BaSO$_4$ and SrSO$_4$ precipitated as a function of pore volume of seawater injected (after Haarberg *et al*[17])

Pore volume of seawater injected (%)	BaSO$_4$ Precipitate (g/m^3)	SrSO$_4$ Precipitate (g/m^3)
0	0.0	0.0
10	71.0	0.0
20	65.0	0.0
30	58.0	45.0
40	48.0	68.0
50	42.0	58.0
60	32.0	26.0
70	25.0	0.0
80	18.0	0.0
90	10.0	0.0
100	0.0	0.0

Table 2. Fluid and reservoir base case properties[7, 12] used as input in the scale prediction model

Pay thickness (h)	26m
Initial permeability	0.5922E-13m^2 (60mD)
Initial porosity	0.04
Reservoir pressure	36600kpa
Bottom hole pressure	22060kpa
Reservoir temperature	353K (80C)
Formation volume factor	0.254
Viscosity	0.0007Pa-s
Connate water saturation	0.2

Table 3. Past investigators and their models up till date

Investigators	Equations	Comments
Gruesbeck and Collins (1982)	$\dfrac{K_P}{K_i} = \exp\left[-a(\sigma^P_s)_P^4\right]$	Empirical for pluggable paths, a is constant
	$\dfrac{K_{np}}{K_i} = \left[1 - b(\sigma^P_s)_{np}\right]^{-1}$	Empirical for non-pluggable paths, b is constant
Soo et al (1986)	$\dfrac{K}{K_i} = 1 - \beta \dfrac{\delta}{\phi_0}$	β is the average flow restriction parameter
Wojtanowicz et al. (1987,1988)	$\dfrac{K}{K_i} = (1 - C_1 t)^2$	Gradual pore blocking $C_1 = \dfrac{\rho_p^f \alpha}{\rho_p L}$
	$\dfrac{K}{K_i} = 1 - C_2 t$	Single pore blocking (screening) $C_2 = \dfrac{6Q\rho_p^f C_s A^2}{\pi C_1 d^3 \rho_p}$
	$\dfrac{K}{K_i} = \dfrac{1}{1 + C_3 t}$	Cake forming (straining) C_3 is constant

Table 3. (Continued)

Investigators	Equations	Comments
Khilar and fogler et al. (1982,1987)	$\dfrac{K}{K_i}\left[1 - B\dfrac{\sigma_s^p}{(\sigma_s^{*p})_i}\right]^2$	Theoretical model based on the Hagen-Poiseuille flow through pore throat, B is a parameter dependent on the characteristic
Civian et al (1990)	$\dfrac{K}{K_i} = \left(\dfrac{\phi}{\phi_i}\right)^3$	Power law assumption
Rachon et al (1996)	$\dfrac{K}{K_i} = \exp[\beta(\phi - \phi_i)]$	Linear relationship between the logarithm of permeability and porosity β is relationship coefficient.
Civian (2001)	$\dfrac{K}{K_o} = 1 - \dfrac{abq^2 t}{r^2}$	Permeability – porosity relationship based on deposition.
Moghadasi et al. (2006a, 2006b) and Kalantari et al. (2006)	$\dfrac{K}{K_i} = (1 - MC_i.t)^3$	Modified Civian et al permeability-porosity correlation based on solid particle precipitation.

Table 4. Our models

Fadairo and Omole	$\dfrac{K}{K_i} = \left[1 - F\left(\dfrac{dC}{dP}\right)_T .t\right]^3$	Modified Civian et al permeability-porosity correlation based on fractional change in solid particle concentration due to pressure drop

DISCUSSION OF RESULTS

The permeability variation due to precipitation of $BaSO_4$ and $SrSO_4$ scale for different pore volumes of seawater injected in a reservoir, during production are shown in Figure 1. From the figure it can be seen that at low pore volume of seawater injected in the range of 0% to 10%, permeability decline due to the precipitation of $BaSO_4$ scale was dominant while there was no visible change in permeability due to $SrSO_4$ scale until pore volume of 20% of seawater was injected. However, at higher pore volume of 40% of seawater injected permeability decline due to the formation of $SrSO_4$ scale is greater than that due to the formation of $BaSO_4$ scale. In other words, the degree of formation damage caused by $BaSO_4$ scale is higher at lower pore volume of seawater injected while there is a higher permeability damage caused by $SrSO_4$ scale at higher pore volume of seawater injected. At every radial distance away from the well bore, the worst scenario of formation damage occurred at low pore volume of 10% of seawater injected. At ratio of 10% seawater to 90% formation water, rapid blocking of the pores was observed (high formation damage) even at relatively low water production rate (Haarberg et al., 1992).

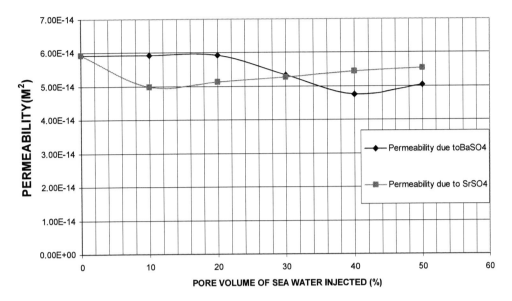

Figure 1. Permeability Variation as a Function of Pore Volume of Seawater Injected when B_aSO_4, and S_rSO_4 Scale Co-precipitated.

Higher rate of deposition occurs at a closer radial distance to the well bore and the degree of permeability damage decreases greatly with increasing radial distance as reported in figure 2. The locations with highly positive skin factor are most likely to experience significant flow impairment by deposited scale.

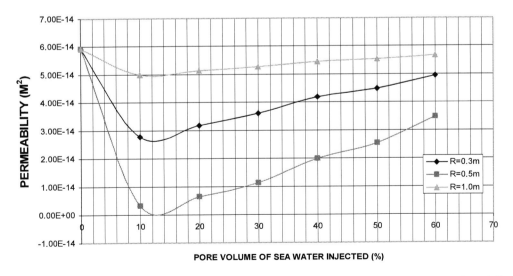

Figure 2. Permeability Induced by B_aSO_4 Scale as a Function of Pore Volume of Seawater Injected @ Different Radial Distances.

Figure 3 shows the effect of flow rate on the skin factor as the pore volume of seawater injected increased. High positive skin occurred earlier for the high flow rate case than the lower flow rate case.

Therefore reduction in flow rate will generally increase the production time of a well prior to significant flow impairment by deposited scale, resulting in less positive skin factor. At a given flow rate, the degree of formation damage induced by scale around the well bore significantly increased by high pressure gradient near the well bore.

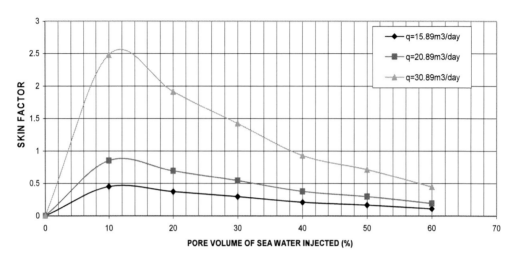

Figure 3. Effect of Flow Rate on Skin Factor Induced by B_aSO_4 @Different Pore Volume Injected.

The previous models under- estimated the degree of permeability damage induced by oilfield scale. Figure 4 compares the results obtained from the previous models with our modified models.

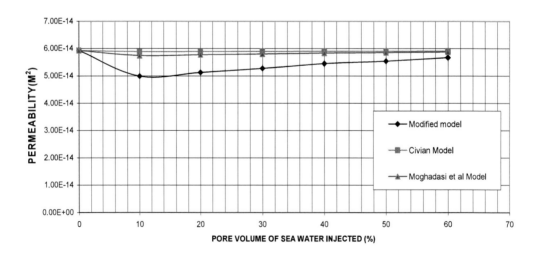

Figure 4. Comparison of Permeability Variation as a Function of Pore Volume of Seawater Injected when B_aSO_4 Deposited Around the Well Bore.

CONCLUSION

The following conclusions were drawn from the result of this study

- Permeability decline caused by scale deposition in the formation ranged from less than 30% to more than 90% of the initial permeability value depending on solid salts concentration, initial permeability and water injection period. High rate of permeability damage occurred under condition of high salt concentration and in low value of formation permeability.
- At a given water injection rate, the degree of formation damage around the well may be significantly reduced by decreasing the pressure gradient near the well bore.
- Reduction in water injection rate generally decreases the degree of formation damage around the well bore and prolongs the injection time prior to significant pores blockage by deposited scale.

RECOMMENDATION

The presented model can predict the degree of permeability damage induced by oilfield scales per percentage of seawater injected, prior to production operation and water flooding scheme. Hence

- Flow rate of produced water should be relatively low, to prevent high deposition of scale that result to well damage
- The analytical models could be introduced into existing scaling prediction software like multiSCALE 6.1 to account for effect of key operational and reservoir parameters on the degree of permeability damage induced by scale deposition around well bore.
- The models could be used for diagnosis, evaluation and simulation of reservoir permeability damage during production operations and water flooding scheme.

ACKNOWLEDGMENT

We wish to thank Petroleum Technology Development Fund (PTDF) for their financial support in carrying out this research work.

REFERENCES

Atkinson, G., Raju, K. and Howell, R.D. (1991) *"The Thermodynamics of Scale Prediction"*, SPE 21021, pp.209–215, presented at the Society of Petroleum Engineers international Symposium on Oilfield chemistry Anaheim, Canada.

Bedrikovetsky, P.G., Gladstone, P.M., Lope, Jr., Rosario, F.F., Silva, M.F., Bezerra, M.C. and Lima, E.A. (2003) *Oilfield Scaling Part I: Mathematical and Laboratory Modeling*,

SPE 81127, presented in SPE American and Caribbean Petroleum Engineering Conference in Port of Spain Trinidad West Indies.

Chang, F.F. and Civan, F (1996) "Practical model for chemically induced formation damage" *Journal of Petroleum Science and Engineering*, Vol. 17, pp.123–137.

Civan, F. (2001) *Modeling Well Performance under Non Equilibrium Deposition Condition*, SPE 67234, presented at SPE production and Operations Symposium Oklahoma, USA.

Civan, F., Knapp, R.M. and Ohen, H.A. (1989) 'Alteration of permeability by fine particle processes', *Journal of Petroleum Science and Engineering*, Vol. 3, Nos. 1, 2, pp.65–79.

Civan, F., Knapp, R.M. and Ohen, H.A. (1990) 'Alteration of permeability by fine particle processes', *Journal of Petroleum Science and Engineering*, Vol. 3, pp.65–79.

Crabtee, M., Eshinger, D., Fletches, P., Johnson, A. and King, G. (1999) *Fighting Scale – Removal and Prevention*, Oilfield Review Schlumberger, Autumn.

Fadairo, A.S.A. (2004) *Prediction Scale Build up Rate around The Well bore (Nigeria)*, MSc Thesis, Department of Petroleum Engineering, University of Ibadan, Nigeria.

Frank, F., Chang and Civan, F. (1991) *Modeling of Formation Damage Due to Physical and Chemical Interaction between Fluid and Reservoir Rock*, SPE 22856. SPE Annual Technical Conference and Exhibition, Dallas, Texas.

Gruesbeck and Collins (1982). "Entrainment and Deposition of Fine Particle in Porous Media" SPE paper 8430 SPE Journal. Volume 22, Number 6.

Haarberg, T., Selm, I., Granbakken, D.B., Østvold, T., Read, P. and Schmidt, T. (1992) *Scale Formation in Reservoir and Production Equipment during Oil Recovery II" equilibrium Model*, SPE Journal Production Engineering, Volume 7, Number 1.

Kalantari, A., Moghadasi, J. and Gholami, V. (2006) *A New Model to Describe Particle Movement and Deposition in Porous Media*, SPE 103666, presented at First International Oil Conference and Exhibition in Mexico.

Khilar and Fogler et al. (1987). "Colloidally Induced Fines Migration", Reviews in Chemical Engineering, Vol. 4 pp 41.

Moghadasi, J., Müller Steinhagen H., Jamialahmadi, M. and Sharif, A. (2005) 'Model study on the kinetics of oil field formation damage due to salt precipitation from injection', *Journal of Petroleum Science and Engineering*, Vol. 46, No. 4, 30, pp.299–315.

Moghadasi, J., Sharif, A., Kalantari, A.M. and Motaie, E. (2006a) *A New Model to Describe Particle Movement and Deposition in Porous Media*, SPE 99391, presented at 15th SPE Europe Conference and Exhibition, Vienna, Austria.

Moghadasi, J., Kalantari, A. and Gholami, V. (2006b) *Formation Damage due to Asphaltene precipitation resulting From CO2 Gas Injection in Iranian Carbonate Reservoir*, SPE 99631. SPE Europec/EAGE Annual Conference and Exhibition, Vienna, Austria.

Oddo, J.E. and Tomson, M.B. (1994) 'Why scale forms in the oilfield and method of predict it', *SPE 21710*, SPE Journal Production and Facilities, Volume 9, Number 1.

Omole, O. and Fadairo, A.S.A. (2008) *"Prediction of Scale Build up Rate around the Well Bore"* (Accepted for Publication in the J. Petroleum Science and Engineering).

Omole, O. and Osoba, J.S. (1984) 'Carbon dioxide – dolomite interactions during CO2 flooding', *Proceeding of the 34th CIM Annual Technical Meeting*, Nigeria pp 87-95.

Rachon, J., Creusot, M.R., and Rivet P., (1996) "Water Quality for Water Injection Wells" SPE 31122 presented at SPE Formation Damage Control Symposium, Lafayette, USA.

Robert, B.E (1997) *Effect of Sulphur Deposition on Gas Well Performance*, SPE 36707. SPE Annual Technical Conference and Exhibition in Denver, Colorado.

Soo, H., and Radke Clayton, J. (1986) "The Flow Mechanism of Dilute Stable Emulsion in porous Media" Industrial Eng. Chem. Fund, No 23, pp 324-347.

Wojtanowicz, A.K., Krillov, Z. and Langlinais, J.P., (1988) "Study of Effect of Pore Blocking Mechanism on Formation Damage" SPE 16233 presented at SPE Production Operations. Oklahoma City, Oklahoma.

In: New Developments in Sustainable Petroleum Engineering
Editor: Rafiq Islam
ISBN: 978-1-61324-159-2
© 2012 Nova Science Publishers, Inc.

Chapter 6

REVIEW OF THE OPTIMIZATION TECHNIQUES IN GROUNDWATER MONITORING NETWORK DESIGN FOR PETROLEUM CONTAMINANT DETECTION

Abdolnabi Abdeh-Kolahchi[*], Mysore G. Satish, Chefi Ketata and M. Rafiqul Islam
Department of Civil Engineering, Dalhousie University,
Halifax, Nova Scotia, Canada

ABSTRACT

In order to detect petroleum contaminants in groundwater, it is paramount to design a proper monitoring network. Groundwater monitoring network design has advanced in recent years with the Genetic Algorithms optimisation techniques to address and achieve the optimal management strategy. This paper presents and describes various optimisation techniques, especially the state of the art robust Genetic Algorithms and their application in groundwater management focused on groundwater monitoring network. This review describes the Optimisation of groundwater management has become an active area of research in the last several years because of its ability to reduce cost substantially. It is the purpose of this paper to provide a comparison of a variety of optimisation methods on a groundwater problem due to petroleum contamination.

Keywords: petroleum contaminants, genetic algorithms, optimisation, groundwater monitoring network design.

1. INTRODUCTION

Optimization is a mathematical process that, given a set of input variables, minimizes or maximizes an objective function. The main purpose of this paper is to introduce and

---
[*] Corresponding author: kolahchi@dal.ca

demonstrate the capability of optimization techniques used to tackle groundwater pollution due to petroleum contamination for a better groundwater quality. Often groundwater problems are difficult to solve using traditional gradient-based techniques, as these problems are nonlinear, nonconvex, and discontinuous.

Many issues related to water resources require solving optimization problems. These include groundwater management, reservoir optimization, remediation, parameters estimation and calibration, monitoring networks, sampling strategy, and many other issues. Traditionally, optimization problems were solved using linear and nonlinear optimization techniques, which normally assume that the minimized function (objective function) is known in analytical form and that it has a single minimum (generally it will be assumed that the optimization problem is a minimization problem).

In practice however, there are many problems that cannot be described analytically and many objective functions have multiple extrema. In these cases it is necessary to use Global Optimization (GO) because the traditional optimization methods are not applicable, and cannot guarantee an optimal solution. Among the various global optimization techniques, Genetic Algorithm (GA) has become popular during recent years (Goldberg 1989).

The main purpose of this paper is to introduce and demonstrate the capability of optimization techniques used to tackle groundwater problems, especially those associated with cost. The limitation of traditional optimization techniques has motivated this researcher to seek an alternative which is sophisticated and has advantages over the traditional optimization techniques.

The global optimization methods are capable of solving large and highly nonlinear real world field problems. GA is the best of these methods, considered to be a robust technique for solving groundwater optimization problems. GA is an optimization technique that follows the biological processes of natural selection and evolution. A GA operates on a population of decision variable sets. Through the application of three specialized genetic operators: selection, crossover, and mutation, a GA population evolves toward an optimal solution. The following sections briefly describe optimization techniques used in groundwater modeling and their application in groundwater management.

The goal of a formal mathematical optimization-based groundwater management model is to achieve a specified objective within the various limiting restrictions. This paper discusses the applications of various optimization techniques including Genetic Algorithms and their applications and proposes a solution for groundwater monitoring network design.

2. TRADITIONAL OPTIMIZATION APPROACH

Traditional or gradient based methods which also known as the steepest descent methods, are considered basic in unconstrained optimization. They require only information about first order derivatives and hence are very attractive because of their limited computational cost and minimal storage requirements. The rate of convergence is usually very poor. Important characteristics of a function such as minimum, maximum, and direction of gradient could be revealed from the derivate of the function. These characteristics are used as tools for gradient-based nonlinear optimization algorithms.

Gradient-based methods are designed to locate a local minimum of the problem. These algorithms could be used to solve the problem that involves nonlinear methods in both objective function and constraints. These methods start from an arbitrary point and move, through successive iterations, to a point where the convergence criteria are met. If the objective function involves a convex problem, the optimization solution is the local minimum as well as the global minimum. However if the objective function is not convex, then there is no guarantee that the local solution found is also the global solution. The starting point is to use the main rules in determining which local solution will be found for nonconvex problems. If there exist only a few local solutions, the limitation should not be very significant.

2.1. Linear Programming

Groundwater linear optimization is applicable when the aquifer model, objective function, and constraints are linear with respect to stress and hydraulic head. For example the objective function can relate to the pumping rate in this method. Linear programming has been applied in groundwater optimization problems. Although the linear optimization method is computationally efficient and is used in several groundwater flow models, it is limited to a confined aquifer and cannot deal with the transport model effectively.

2.2. Nonlinear Programming

Nonlinear Programming (NLP) in groundwater problems can be used in cases of nonlinearity in both aquifer properties and in objective function. In groundwater management models, nonlinearity may arise due to flow and transport characteristics of the governing equation, cost function, management model, and constraints. Nonlinear programming can overcome the problems due to nonlinearity by employing gradient-based algorithms to adjust decision variables so as to optimize the objective function of management models.

Several researchers have used nonlinear optimization to groundwater applications such as aquifer cleanup problems. However, nonlinear programming techniques cannot guarantee global optimality when applied to large nonconvex problems.

2.3. Differential Dynamic Programming

Differential Dynamic Programming (DDP) is efficient for a large number of management periods and is used to find an optimal solution for a discrete time control problem. DDP methods decompose multistage decision problems to a sequence of single-stage decision problems. The advantage of using DDP is that it can deal with discrete variables, non-convex, non-continuous, and complex functions. In addition DDP significantly reduces the dimensionality difficulties associated with non-linear dynamic groundwater management problems.

The NLP and DDP methods are also called gradient methods where the derivatives of the objective function with respect to decision and state variables are evaluated. Although DDP is computationally efficient, it has limitations and suffers under highly non-linear and complex objective functions. As a result, it is possible to get trapped in a local solution and fail to find the optimal global solution. In addition, they face numerical instability and convergence problems.

2.4. Mixed Integer Linear Programming and Mixed Integer Nonlinear Programming

Mixed Integer Linear Programming (MILP) and Mixed Integer Nonlinear Programming (MINLP) are applied when discrete decision variables (e.g., well location and fixed cost) are involved. Willis (1976, 1979) applied MILP and MINLP methods. When the groundwater management problems dealing with questions like the number of wells, locations, off/on, it is recommended to use this technique.

McKinney and Lin (1995) developed a mixed-integer nonlinear programming model for an optimal aquifer remediation design model to find the minimum cost design of a pump-and-treat aquifer remediation system. The MINLP includes the discontinuous fixed costs of system construction and installation, as well as operation and maintenance.

2.5. Limitations of the Traditional Method

The application of conventional optimization algorithms such as linear and nonlinear programming is difficult due to the discontinuity of the fixed cost function in the objective function and the combinatorial nature of assigning discrete well locations. Groundwater integer programming and/or discrete dynamic programming is suffering from large computational load caused by varying pumping rates over time. In addition, these techniques cannot guarantee global optimal solution, and suffer from highly nonlinear and complex objective function code segment. They can be potentially trapped in local optimal solutions. Furthermore, they show difficulty with solving multi-optimal solution problems. They can display numerical instability and convergence problems.

Some complexity of a problem for these methods lies in the complexity of the solution space that must be searched through. This complexity arises due to:

- Size of the problem domain.
- Nonlinear interactions between various elements.
- Domain constraints dynamic performance measurement, and many independent and codependent elements.
- Incomplete, uncertain, and imprecise information.

3. GLOBAL OPTIMIZATION

The limitations of mathematical programming and nonlinearity in phenomena and processes have motivated researchers to use alternative optimization techniques known as global optimization. Using the global optimization methods could eliminate the limitation of gradient methods for nonconvex problems. These methods are able to avoid trapping in local minimization. They are suitable for discontinuous, noisy, and multiple objective problems. In the case of a complex nonlinear system description, the associated decision model may, and frequently will, have multiple locally-optimal solutions. In the most realistic cases, the number of such local solutions is unknown, and the quality of local and global solutions may differ substantially. Therefore the rules of standard optimization strategies are not directly applicable.

The objective of global optimization (GO) is to determine the absolute best solution and corresponding optimum value in complex models. Global Optimization addresses the computational issues associated with the characteristics of nonconvex functions. The aim of Global Optimization is to find the best solution of constrained optimization problems, which may have various local optima. The GO approach has a particular advantage in problems where other optimization techniques have difficulties due to the existence of multiple extrema and/or difficulties in defining functions analytically (Solomatine, 1998).

Since 1960, various GO techniques have been successfully used and recently the accessibility of GO has increased enormously. In civil engineering, the GA became very popular toward the end of the 1980s and allowed many researchers and practitioners to use a relatively simple and effective optimization technique. One of the reasons that the GA has been chosen to solve such problems is that it is able to solve the nonconvex, discrete, and discontinuous problems (Goldberg, 1989) that arise frequently in these applications. A considerable number of publications related to water resources are devoted to GA usage.

In the 1990's a new class of optimization methods based on heuristic search techniques have emerged. There are various possible algorithms applicable to global optimization problems. Three major representatives of these optimization methods are Simulated Annealing (SA), Genetic Algorithm (GA), and Tabu Search (TS).

The most popular of these is GA technique. These methods are able to identify the global or near global optimal, and gradient-free solution. These recent methods all have a high degree of searching ability. Recently they have been applied to solve optimization problems in groundwater management, since they are able to obtain a globally optimal solution, determine the best management strategy, efficiently handle discrete decision variables such as well location, and easily link to flow and transport simulation models (Zheng et al, 2003).

These optimization techniques have been referred to as Global Optimization (GO) methods because of their ability to identify the global optimum from a complex problem. For example genetic algorithms earn (used) certain natural systems, such as biological evolution to identify the optimal solution, instead of being guided by the gradients of the objective function. These state of the art methods all have a high degree of searching ability. Global optimization methods generally require intensive computational efforts. In spite of this, however, global optimization methods are increasingly being used to solve groundwater management problems. The objective of global optimization (GO) is to determine the absolute best solution and corresponding optimum value in complex models. Global optimization addresses the computation issues associated with the characteristics of nonconvex functions.

Recently the combination of Gentic algorithms and geostatiscis application in groundwater management reported. Geostatistics is practical techniques for the estimation of spatial function from sparse data (Kitanidis 1997) and provides minimum error estimates of a phenomenon at unsampled locations using linear combinations of samples values. In addition to providing the expected value of a phenomenon, the estimation variance is computed that represents the uncertainty of estimates at unsampled locations. Estimation variances can be computed both locally and globally. A local estimation variance accounts for the uncertainty of an estimate at a single unsampled location, whereas a global estimation variance represents the uncertainty accumulated over several unsampled locations within a subset of the interpolation domain (Reed 2002).

4. Genetic Algorithm

Genetic algorithms (GA) are based on the idea of modelling a search process based on natural evolution. It uses random elements to avoid local minima and uses terminology from biology and genetics. GAs gained a lot of popularity as a general-purpose optimization algorithm. Such techniques do not require computation of derivatives as used in more traditional optimization methods. In addition they can be applied in groundwater modelling problems. Unlike gradient-based methods, the GA uses a random search procedure inspired by biological evolution and cross-breeding trial design, allowing only the fittest to survive and propagate to successive generations. This method uses heuristic rules to identify the best solution, and to search for solutions of complex problems using an analogy between optimization and natural selection.

The use of the GA was first suggested by Holland (1975), who based his research on a survival-of-the-fittest rule. Since then, the GA has been used in many disciplines. The basic idea is to evolve a population of candidate solutions to the given problem, using operators inspired by natural selection. Goldberg (1989) presents a comprehensive introduction to the GA as does Davis (1991) who reviews many important applications of the GA. When using the GA to solve an optimization problem, decision variable values are encoded as substring of binary digits or real numbers.

As previously stated, the GA is based on the Darwinist theory (survival of the fittest), where the strongest (or any selected) offspring in a generation are more likely to survive and reproduce. The method starts with a number of possible solutions, referred to as the initial population. They are randomly selected within the predetermined lower and upper boundaries of each model parameter to be optimized. Each of the possible solutions in the initial population is referred to as an individual, typically encoded as a binary string. For each individual, the objective function is evaluated. During the course of the research, new generations of individuals are reproduced from the old generations through random selection, crossover, and mutation based on certain probabilistic rules. The selection is in favour of those interim solutions with lower objective function values in a minimization problem. Gradually, the population will evolve toward the optimal solution. At the core of the GA are three basic operators: selection, crossover, and mutation. GA procedures detailed in Goldberg (1989), Sen and Stoffa (1995), and Obitko (1998).

The success and performance of the GA depends on several parameters: population size, number of generations, and the probabilities of crossover and mutation. It has been suggested that good GA performance requires the choice of high crossover and low mutation probabilities and good population size. This also confirm by (Abdeh-Kolahchi, 2006) using sensitivities analysis of GA parameters on solution of objective function of groundwater monitoring network design. Some researchers have suggested that the probability of mutations required is inversely proportional to population size and this would be enough to prevent the search from looking onto a local optimum.

GAs have been used for many water resources case studies. Applications include groundwater remediation design, optimal reservoir system operation, calibrating rainfall-runoff models, remediation policy selection, and solving multiple-objective groundwater contaminant problems. Many of these models are nonlinear due to the behaviour of groundwater processes and management models which can solve by GA.

5. GA ADVANTAGES

Genetic algorithm methods are robust and can provide more than one solution, i.e., in addition to the best solution. They can search several near-optimal solutions. If the best solution is difficult to implement, the next best solution can be used. This is very practical and useful especially for groundwater management situation. Genetic algorithms are random search algorithms that remove the shortcomings of calculus-based methods involving derivatives and multi-peaks. They can search for an optimal solution from the entire population not just from a single point. They perform well in large search spaces, i.e. they cover broad searches as well as local searches. They can solve discrete, nonconvex, and discontinuous problems.

Genetic algorithms ensure that the best solution will evolve towards the solution of objective function. They can optimize functions with either continuous or discrete parameters. They are proficient in dealing with large number of parameters. They can handle complex functions. They are flexible enough to easily jump out of the local optimal solution, and adept at providing a list of optimal solutions, rather than stopping at one. Such techniques do not require computation of derivatives as required in more traditional optimization methods and can be applied in groundwater modelling problems.

The only problem with genetic algorithms is that they do not provide an explicit method for handling constraints. Instead of explicitly considering constraints, penalty function terms must be added to the objective function.

6. SIMULATED ANNEALING

The motivation for simulated annealing (SA) comes from a similarity between the physical annealing process of solids and optimization problems. A solid is annealed by increasing its temperature so that its molecules are highly mobile, followed by slow cooling to force them into the low energy state of a crystalline lattice. In optimization, the objective function represents the energy in the thermodynamic process, while the optimal solution corresponds to the crystal state. Through stochastic relaxation, SA has the ability to escape from local minimum points. The essential feature of SA lies in allowing the procedure to move both 'uphill' and 'downhill' (Zheng and Wang, 2003).

Simulated Annealing (SA) introduced by Kirkpatrick et al., (1983) and its application in groundwater optimization problems can be found in Dougherty and Marryott (1991),

Marryott et al. (1993), Mauldon et al. (1993), Rizzo and Dougherty (1996), and Wang and Zheng (1998).

7. LITERATURE REVIEW OF GROUNDWATER OPTIMIZATION

Carrera et al. (1984) developed a method, based on nonlinear programming and a special branch and bound technique for selecting optimal locations from a discrete set of possible measurement points. The method was applied to find the optimal location of measurement points to estimate the fluoride concentration in groundwater.

Gorelick, et al., (1984) developed a simulation management model that combines finite element groundwater flow and contaminant transport simulation with nonlinear optimization. They used constraints on hydraulic heads, stresses, gradients, contaminant concentration, and fluxes distributed over space and time. This method is capable of determining well locations, as well as pumping and injection rates for groundwater pump and treat systems.

Atwood and Gorelick (1985) developed a two-stage planning procedure that successfully selected the best wells and their optimal pumping and recharge schedules to contain a plume, while a well or system of wells within the plume removed the contaminated water. After the simulation of flow and contaminant transient, in stage II, a linear program, including a groundwater flow model as part of the set of constraints, determined the optimal well selection and their optimal pumping and recharge schedules by minimizing total pumping and recharge. The simulation–management model eliminates wells far from the plume perimeter and activates wells near the perimeter as the plume decreases in size.

Jones, et al. (1987) applied Differential Dynamic Programming (DDP) for optimal control of large-scale nonlinear groundwater hydraulics management.

Ahlfeld, et al. (1988) addressed the problem of designing contaminated groundwater remediation systems using hydraulic control. The proposed model used nonlinear optimization formulations to design a process to determine the location and pump rates of injection and extraction wells in an aquifer cleanup system. They seek a pumping rate that removes the maximum amount of contaminant over a fixed time period which reduces the contaminant concentration to specified levels by the end of a fixed time period at least cost.

Dougherty and Marryott (1991) introduced and successfully applied the Simulated Annealing (SA) optimization to groundwater management problems in four examples using a hypothetical site. The first three are specific examples of the well rate problem while the fourth combines the well rate problem with a slurry wall.

Chang, et al., (1992) developed a method to compute optimal time-varying pumping rates for the remediation of contaminated groundwater. They combined a pollutant transport model with a constrained optimal control algorithm. The transport model simulates the unsteady fluid flow and transient contaminant advection-dispersion in a two-dimensional confined aquifer. The constrained optimal control algorithm employs a hyperbolic penalty function. They concluded that the optimal constant pumping rates were 75% more expensive than the optimal time-varying pumping rates.

Culver and Shoemaker (1992) developed a successive approximation linear quadratic regulator (SALQR) method with management periods, which was combined with a finite element groundwater flow and transport simulation model to determine optimal time-varying groundwater pump-and-treat remediation policies. The optimal policies, including the number and locations of wells, changed significantly with the number of management periods. Complexity analysis revealed that the SALQR algorithm with management periods could significantly reduce the computational requirements for non-steady optimization of groundwater remediation and other management applications.

Marryott et al. (1993) used simulated annealing to analyze alternate design strategies for groundwater remediation at a contaminated field site. The simulated annealing optimization algorithm was combined with a field-scale flow and transport simulation model for an unconfined aquifer to determine nearly optimal pumping schedules for a pump-and-treat remediation system at a proposed Superfund site in central California. A series of demonstration problems were presented using two different optimization formulations. The results of these experiments indicated that the method could be applied to groundwater management problems. The computational expense of simulated annealing is large, yet comparable to other nonlinear optimization techniques. An empirical strategy was provided for selecting and adjusting the parameters necessary for successful optimization.

Ahlfeld and Heidari (1994) reviewed the optimal hydraulic-control approach to field problems. The deterministic hydraulic control problem was contrasted with more sophisticated techniques that incorporated contaminant transport and uncertainty in aquifer parameters. The optimal hydraulic-control problem for groundwater systems was defined in terms of controls on the groundwater system and typical design criteria. Commonly used linear formulations of the control problem were described in detail.

Meyer et al. (1994) presented a method that incorporated system uncertainty in monitoring network design and provided network alternatives that were superior with respect to several objectives. The random inputs to the simulation were the hydraulic conductivity field and the contaminant source location. The design objectives considered were:

- Minimize the number of monitoring wells.
- Maximize the probability of detecting a contaminant leak from landfill.
- Minimize the expected area of contamination at the time of detection.

The network design problem was formulated as a multi-objective, integer-programming problem, and was solved using simulated annealing.

McKinney and Lin (1994) combined groundwater simulation models with GA optimization to solve three groundwater management problems: maximum pumping from an aquifer, minimum cost water supply development, and minimum cost aquifer remediation. The results showed that genetic algorithms could effectively and efficiently be used to obtain globally optimal solutions to these groundwater management problems. Also using a GA to compute the starting point for a nonlinear gradient-based optimization algorithm provided significant advantages and located solutions that are approximately globally optimal. More complicated problems, such as transient pumping and multiphase remediation, can be formulated and solved using this method. The computational time required for the solution of genetic algorithm groundwater management models increases with the complexity of the problem.

Ritzel et al. (1994) applied GA to a multiple-objective groundwater pollution containment problem. The problem involved finding the set of optimal solutions on the trade-off curve between the reliability and cost of a hydraulic containment system. The decision variables were the number and location of wells to install, and the amount water to pump from each well. The results suggested that a GA performed better than mathematical programming for nonlinear and mixed-integer nonlinear problems.

Rogers and Dowla (1994) presented and used a combination of Artificial Neural Networks (ANNs) and GA to optimize aquifer remediation. In this approach, optimal management solutions were found by:

- Training the ANN to predict the outcome of the flow and transport code.
- Using the trained ANN to search through many pumping realizations to find an optimal one for successful remediation.

The optimization technique involved minimizing the cost, cleanup time, and the contaminant. The results suggested that the combination of ANN and GA had advantages over conventional techniques. They found that this combination involved less computational equipment and more flexibility than mathematical programming methods.

Taghavi et al. (1994) developed a management model that used a nonlinear programming approach to solve the optimal control problem of managing the generation and disposal of agricultural and dairy waste. The permissible controls were subject to physical, economic, and social constraints. This method is different from other combined simulation/optimization approaches, since it includes the dynamic response of the system as an explicit part of the optimization.

Cieniawski et al. (1995) investigate a method of optimization using GA, which allowed consideration of the two objectives, maximizing reliability and minimizing contaminated area at the time of first detection, separately yet simultaneously. Four different codings of GAs are investigated, and their performance in generating the multiobjective trade-off curve is evaluated for the groundwater monitoring problem. The study also compared the results of GA and simulated annealing.

McKinney and Lin (1995) developed a design model for optimal aquifer remediation employing a nonlinear programming find the minimum cost design for a pump-and-treat system. They used a mixed-integer nonlinear programming model. Results showed that a combined well field and treatment process model that included fixed costs had a significant impact on the design and cost of aquifer remediation systems, reducing system costs by using fewer and larger flow rate wells.

Takahashi and Peralta (1995) developed optimal groundwater yield pumping strategies for a complex multilayer aquifer. They used linear response matrix and embedding (EM) simulation/optimization technique for piecewise-linear constraints for evapotranspiration, discharge from flowing wells, drain discharge, and vertical interlayer flow reduction due to desaturation of a confined aquifer.

Datta and Dhiman (1996) applied mixed integer programming in groundwater monitoring network design. They used the objective of maximizing the probability of detecting contamination at locations where a specified standard concentration was exceeded.

Harrouni et al. (1996) used GA in to solved two optimization problems pumping management problem in a homogeneous aquifer and parameter estimation in a heterogeneous aquifer. The initial findings regarding the effectiveness of the GA were successful.

Culver and Shoemaker (1997) used a dynamic optimal control algorithm for groundwater remediation, which was extended to incorporate treatment facility capital costs as a function of the peak operating rate. In cooperating of the capital costs of the treatment facility had the greatest impact on design when the pumping policies were changed at intervals of six months or less. For longer management periods, inclusion of treatment facility capital costs had little effect on the selected optimal policies. This work demonstrated that capital treatment costs might significantly impact a dynamic management policy and that these capital costs should be explicitly incorporated into a dynamic management model.

Huang and Mayer (1997) presented an optimization formulation for dynamic groundwater remediation management, that simultaneously used well locations and the corresponding pumping rates as the decision variables. GA was applied to search for optimal pumping rates and the discrete spacing of well locations. The optimization model was used in hypothetical, three-dimensional, contaminated aquifer systems with homogeneous and heterogeneous porous media properties. The well location search path and convergence behaviour indicated that the GA was an effective alternative solution scheme and that well location optimization was more important than pumping rate optimization.

Kwanyuen and Fontane (1998) solved a groundwater development planning problem considering the minimization of both fixed installation and variable operation costs, formulated using the response matrix method. This mixed integer, nonlinear problem was solved using penalty coefficient, pseudo-integer, and heuristic branch-and-bound methods.

Minsker and Shoemaker (1998) coupled optimal control and simulation models to select injection and extraction well sites and pumping rates for cost-effective in-situ bioremediation design. Dynamic optimal control is the most efficient for time-varying problems, but has disadvantages including difficulty in managing discrete or discontinuous cost functions such as well installation costs. To overcome this disadvantage and to improve performance, they proposed the application of a Hybrid Genetic Algorithm (HGA), which was able to solve complex problems much faster. For this purpose they used a hypothetical site. They identified a number of findings that might be useful for improving in-situ bioremediation design.

Morshed and Kaluarachchi (1998) performed inverse groundwater modelling for parameter estimation and used ANN regression analysis and ANN-GA optimization techniques. The critical parameters were the grain size distribution index and the saturated hydraulic conductivity of water. Using limited monitoring and recovery well data under homogeneous and heterogeneous conditions, the critical parameters were evaluated. The results of the work suggested that ANN regression analysis had limited utility, especially with heterogeneous soils, whereas the ANN-GA optimization could provide superior results with better computational efficiency.

Wang and Zheng (1998) developed a groundwater simulation-optimization based on genetic algorithm for optimal design of groundwater remediation systems under a variety of field conditions. The model is capable of determining optimal time-varying pumping rates under a wide variety of field conditions. The objective function can be highly nonlinear and very general to include drilling, installation, pumping, and treatment costs. The model is applied to a typical two-dimensional pump-and-treat example as well as to a large-scale three-dimensional field problem to determine the minimum pumping needed to contain an existing contaminant plume.

Aly and Peralta (1999a, 1999b) presented and applied a simulation/optimization approach for single and multiple planning period problems in groundwater remediation. They used a GA to find the global optimal solution and incooperation a neural network to model the response surface within the genetic algorithm. They compared the performance of mixed integer, nonlinear programming and a genetic algorithm for several optimization scenarios. As a result of the successes of this approach, Aly and Ruskauff (2002) developed the Adaptive Simulated Annealing Package (ASAP) software to achieve optimal solutions in water resources and environmental management problems. This software is a combination of MODFLOW and MT3D with artificial neural network (ANN) and adaptive simulated annealing (ASA) techniques. This approach can result in large reductions in initial capital and long-term optimization and monitoring costs in a wide range of environmental and water resources projects, including, but not limited to, pump-and-treat groundwater remediation, water resource planning, and risk-based remediation projects. This method was successfully used in several locations with remediation problems.

Zheng and Wang (1999) used tabu search as global optimization, integrated with linear programming to solve remediation design problems. This integrated approach took advantage of the fact that the tabu global optimization technique was most effective for optimizing discrete well location variables, while linear programming is much more efficient for optimizing continuous pumping rate variables. It was also demonstrated that the maximum number of wells allowed in a given design had a significant effect on the total remediation costs. The total remediation costs are nearly doubled when only one well is allowed instead of the optimal number for the test problem.

Yoon and Shoemaker (1999) compared computational performance of eight optimization algorithms used to identify the most cost effective policy for in situ bioremediation of contaminated ground water. Numerical results were obtained for bioremediation of three problems based on two aquifers with time-invariant or time-varying pumping rates. Three major classes of algorithms were considered in the comparison:

(1) Evolutionary algorithms (binary-coded genetic algorithm (BIGA), real-coded genetic algorithm, and de-randomized evolution strategy (DES)).
(2) Direct search methods (Nelder-Mead simplex, modified simplex, and parallel directive search)
(3) Derivative-based optimization methods (implicit filtering for constrained optimization and the successive approximation linear quadratic regulator).

Based on the three problems considered, the successive approximation linear quadratic regulator was the fastest algorithm. No single algorithm was consistently the most accurate in all three problems.

Aksoy and Culver (2000) used GA optimization of pump-and-treat ground water remediation to explore the extent of bias introduced into remediation designs and costs by sorption assumptions. Two optimization formulations were considered for each sorption assumption. In one formulation, the final water quality goal was with respect to the aqueous concentration. In the other, the goal was with respect to the total concentration.

The genetic algorithm is used to determine the values of the decision variables (length of the remediation period), and these values are then used as input into the contaminant transport simulation model to determine the resulting concentration and head values at the end of the remediation period. The optimal designs with equilibrium and kinetic sorption are then compared.

The risk management groundwater model formulation is highly nonlinear and discontinuous. Traditional, derivative-based optimization methods could not be used to solve this model without substantial modification. For this reason, genetic algorithm is a good candidate for the solution algorithm. However, simple genetic algorithms cannot explicitly consider uncertainty. The authors investigated a method called a noisy genetic algorithm, which offered promise as a more computationally efficient method that could be used to consider highly complex forms of uncertainty. Noisy genetic algorithms are simply ordinary genetic algorithms that operate in noisy environments. The 'noise' that exists in certain environments can be defined as any factor that hinders the accurate evaluation of the 'fitness' of a given trial design.

Chan Hilton and Culver (2000) introduced methods for handling constraints when applying a genetic algorithm to optimal ground-water remediation design problems. They compared two methods for constraint handling within the genetic algorithm framework. The first method, the additive penalty method (APM), is a commonly used penalty function approach in which a penalty cost proportional to the total constraints violation is added to the objective function. The second method, the multiplicative penalty method (MPM), multiplies the objective function by a factor proportional to the total constraints violation. These methods are applied to two pump-and-treat design examples.

Morshed and Kaluarachchi (2000) critically reviewed recent research related to the application of GA in solving optimization problems. They identified and explored three areas of potential enhancement to GA. These enhancement methods were fitness reduction method (FRM), search bound sampling method (SBSM), and optimal resource allocation guideline (ORAG). In order to assess these methods, a nonlinear groundwater problem with fixed and variable costs was selected where the corresponding optimal solution using a gradient-based nonlinear programming (NLP) technique was available. The problem was resolved using GA coupled with the enhancement methods. In addition, the GA solutions were compared with the NLP solutions. The new methods increased efficiency, accuracy, and reliability of the solution.

Reed et al. (2000) developed a new methodology for sampling plan design to reduce the costs associated with long-term monitoring of sites with groundwater contamination. The method combined a fate-and-transport model, a plume interpolation, and a GA to identify cost-effective sampling plans that accurately quantified the total mass of dissolved contaminant. The plume interpolation method was considered by using geostatistical approaches.

Reed et al. (2000) discussed the relationships of parameters of GA (selection, recombination, and mutation) and applied in a long-term groundwater monitoring design. These relationships provided a highly efficient method to ensure convergence to near-optimal or optimal solutions. Application of the method in a monitoring design test case identified robust parameter values using only three trial runs.

Smalley et al. (2000) developed a management model that simultaneously predicted risk and proposed cost-effective options for reducing risk to acceptable levels under conditions of uncertainty. The model combined a noisy genetic algorithm with a numerical fate and transport model, and an exposure and risk assessment model. Results from an application to a site from the literature showed that the noisy genetic algorithm was capable of identifying highly reliable designs from a small number of samples, which a significant advantage for computationally intensive groundwater management models. For the site considered, time-dependent costs associated with monitoring and the remedial system were significant, illustrating the potential importance of allowing variable cleanup lengths and a realistic cost function.

Karpouzos et al., (2001) used GA to solve the inverse problem of groundwater flow optimization. First, a multi-population genetic algorithm was developed. It was then tested on two synthetic cases of steady state flow with transmissivity values extending over 4 orders of magnitude. In view of the good quality of the results, it was recommended to use genetic algorithms in inverse groundwater problems.

Yoon and Shoemaker (2001) developed a more efficient GA that was applied to in situ bioremediation of groundwater. The algorithm involved a Real-coded GA (RGA) coupled with two newly developed operators: directive recombination and screened replacement. The numerical results obtained for two bioremediation examples indicated that these operators significantly improved the performance of RGA and that RGA performed much better than the standard binary-coded GA for the groundwater remediation problem.

Recent applications of Differential Dynamic Programming (DDP) presented by Chang and Hsiao (2002) and Hsiao and Chang (2002). They introduced a novel algorithm that integrated a GA and constrained differential dynamic programming (CDDP) to solve a time-varying ground water remediation problem. A GA could easily incorporate the fixed costs associated with the installation of wells. However, using a GA to solve problems for time-varying policies would dramatically increase the computational resources required. Therefore, the CDDP was used to handle the sub-problems associated with time-varying operating costs. A hypothetical case study that incorporated fixed and time-varying operating costs was presented to demonstrate the effectiveness of the proposed algorithm. Simulation results indicated that the fixed costs could significantly influence the number and location of wells, and a notable total cost savings could be realized by applying the novel algorithm.

Hsiao and Chang (2002) built a procedure that integrated GA with constrained differential dynamic programming (CDDP), which calculates optimal solutions for a groundwater resources planning problem while simultaneously considering fixed costs and time-varying pumping rates. The GA determined the number and locations of pumping wells with operating costs, which was then, evaluated using CDDP. They demonstrated that fixed costs associated with installing wells significantly impacted the optimal number and locations of wells.

Maskey et al. (2002) discussed four global optimization (GO) algorithms, including a popular GA and used them to minimize both cleanup time and cleanup cost taking pumping rates and/or well locations as decision variables. Groundwater flow and particle-tracking models (MODFLOW and MODPATH) and a GO tool (GLOBE) were used. Real and hypothetical contaminated aquifers were considered. The results showed that GO techniques could be widely applied in groundwater remediation strategy and planning.

Zheng and Wang (2002) combined groundwater simulation and GA optimization to design a pump and treat remediation system and clean up contamination.

Abdeh-Kolahchi (2006) combined GA with simulation of groundwater contamination to identify the optimal location and exact depth of monitoring well for groundwater monitoring network design. To control, prevent, and remediation groundwater contamination, large number of monitoring well locations is required in a 3-dimensional transient system. This is associated with significant installation, operational and implementation costs. Therefore, a method, which can identify optimal number of monitoring wells and its exact depth, is useful in saving costs and for effective monitoring of the plume concentration and movement.

A state of the art groundwater monitoring network design, which combines groundwater flow and transport results with a Genetic Algorithm (GA) based optimization procedure to identify optimal monitoring well location is developed and implemented for a real world filed by Abdeh-Kolahchi (2006). The proposed sequential network design approach differs from other monitoring network designs by placing the emphasis on maximizing the probability of tracking a transient contamination plume by determining sequential monitoring locations. That study also addresses the issue of enhancing modelling accuracy when the hydrogeologic and hydrochemical data such as contaminant concentration measurement data are sparse. The groundwater flow and contamination simulation results are introduced as input to the optimization model, using Genetic Algorithm (GA) to identify the groundwater optimal monitoring network design, based on several candidate monitoring locations. The groundwater monitoring network design model uses Genetic Algorithms with binary variables representing potential monitoring locations and is capable of finding the global optimal solution to a monitoring network design problem involving 18.4 E+18 solutions.

CONCLUSION

In the last three decades, several groundwater management models have been developed for various applications such as monitoring network design, search for contamination sources, agricultural water supply, conjunctive use of surface and groundwater for quality and quantity, containment and isolation of the contaminant plume in the groundwater, and remediation of contaminate groundwater. In order to solve optimization based groundwater management models, researchers have used various mathematical programming techniques such as linear programming (LP), non-linear programming (NLP), mixed integer programming (MIP), differential dynamic programming (DDP), and global optimization techniques such as genetic algorithms. State of the art optimization techniques in groundwater management especially, the application of genetic algorithms in groundwater monitoring network design is presented in this paper.

The summary of this paper presented in Table 1, which indicates the mathematical optimization techniques and their application in groundwater, with references.

Table 1. Application of optimization methods in groundwater management

Optimization Methods	Authors	Objective
Linear Programming	Atwood and Gorelick (1985)	Best well location and their optimal pumping/recharge
	Ahlfeld and Heidari (1994)	Optimal hydraulic-control
	Takahashi and Peralta (1995)	Optimal groundwater pumping strategies
Non-linear programming	Carrera et al. (1984)	Optimal location
	Gorelick et al. (1984)	Well locations plus pumping and injection rates
	Ahlfeld et al. (1988)	Location and pump rates of injection and extraction for remediation
	Taghavi et al. (1994)	Optimal control problem of managing the generation and disposal of agricultural and dairy waste

Table 1. (Continued)

Optimization Methods	Authors	Objective
Differential Dynamic Programming	Jones et al. (1987)	Optimal control of large-scale groundwater hydraulics management
	Culver and Shoemaker (1997)	Dynamic optimal control
	Chang and Hsiao (2002)	Time-varying ground water remediation problem
	Hsiao and Chang (2002)	Time-varying ground water remediation problem
Mixed Integer Linear/Non-Linear Programming	McKinney and Lin (1995)	Minimum cost design for pump-and-treat system
	Datta and Dhiman (1996)	Detecting contamination
	Kwanyuen and Fontane (1998)	Installation and operation cost
Simulated Annealing (SA)	Dougherty and Marryott (1991)	Well rate problem and slurry
	Marryott et al. (1993)	Optimal pumping schedules for a pump-and-treat remediation system
	Meyer et al. (1994)	Contamination detection
	Aly and Peralta (1999b)	Remediation
	Aly and Ruskauff (2002)	Developed the Adaptive Simulated Annealing Software
	Rizzo and Dougherty (1996)	Multiple management period
Genetic Algorithm (GA)	McKinney and Lin (1994)	Groundwater management problems (pumping rate, cost, and remediation)
	Ritzel et al. (1994)	Multiple-objective groundwater containment problem
	Cieniawski et al. (1995)	Maximizing reliability and minimizing contaminated area at the time of first detection
	Harrouni et al. (1996)	Pumping rate and parameter estimation
	Huang and Mayer (1997)	Pumping rates and the discrete spacing of well locations
	Minsker and Shoemaker (1998)	Select injection/extraction well sites and pumping rates for cost
	Wang and Zhang (1998)	Simulation-optimization using GA
	Aly and Peralta (1999a)	Single and multiple planning period remediation
	Yoon and Shoemaker (1999)	Compare cost effective policy of bioremediation using eight optimization techniques
	Reed et al. (2000)	Cost associated with long term monitoring
	Reed et al. (2000)	Long-term groundwater monitoring design
	Aksoy and Culver (2000)	Pump and treat remediation
	Smalley et al. (2000)	Predicts risk and cost
	Chan Hilton and Culver (2000)	Constraint handling

Optimization Methods	Authors	Objective
	Morshed and Kaluarachchi (2000)	Enhancement of GA
	Karpouzos at el., (2001)	Inverse problem of groundwater
	Yoon and Shoemaker (2001)	Groundwater bioremediation
	Maskey et al. (2002)	Minimization of cleanup time and cost
	Zheng and Wang (2002)	Remediation
	Abdeh-Kolahchi (2006)	Tracking the movement of contamination
Tabu	Zheng and Wang (1996) and (1999)	Parameter identification Remediation cost
Geo-Statistics	ASCE (1990a, 1990b)	Review on Geo-statistics in geo-hydrology
	Rouhani (1985)	Monitoring location
	Rouhani And Hall (1988)	Identify monitoring location
	Cameron and Hunter (2000)	Reducing sampling redundancy
	Reed (2002)	Interpolating concentration
Constrained Optimal Control Algorithm	Chang et al. (1992)	Optimal constant pumping rates vs. optimal time-varying pumping rates
Successive Approximation Linear Quadratic Regulator (SALQR)	Culver and Shoemaker (1992)	Optimal time-varying groundwater pump-and-treat remediation policies
Artificial Neural Network	Rogers and Dowla (1994) Morshed and Kaluarachchi (1998)	Optimize aquifer remediation Parameter estimation

ACKNOWLEDGMENT

The support from the Natural Sciences and Engineering Research Council of Canada (NSERC) Grant 43353 is gratefully acknowledged.

REFERENCES

Abdeh-Kolahchi, A. (2006) Optimal Dynamic Monitoring Network Design For Reliable Tracking of Contaminant Plumes in aquifer, *Civil Engineering*, Dalhousie University.

Ahlfeld, D.P. and Heidari, M. (1994) 'Applications of optimal hydraulic control to ground-water systems', *J. Water Resour. Plan. Manage.*, 120(3), pp. 350–365.

Ahlfeld, D.P., Mulvey, J.M., Pinder, G.F. and Wood, E.F. (1988) 'Contaminated groundwater remediation design using simulation, optimization, and sensitivity theory. 1: Model development', *Water Resour. Res.*, 24(5), pp. 431–441.

Aly, A.H. and Peralta, R.C. (1999a) 'Comparison of a genetic algorithm and mathematical programming to the design of groundwater cleanup systems', *Water Res. Res.*, 35(8), pp. 2415-2426

Aly, A.H. and Peralta, R.C. (1999b) 'Optimal design of aquifer cleanup systems under uncertainty using a neural network and a genetic algorithm', *Water Res. Res.*, 35(8), pp. 2523-2532

Aly, A.H. and Ruskauff, R.C. (2002) 'Adaptive Simulated Annealing Package (ASAP)', http://www.waterstoneinc.com/Home/home_2.html.

Aksoy, A. and Culver, T.B. (2000) 'Effect of sorption assumptions on aquifer remediation designs' Ground Water, 38(2), pp. 200-209.

ASCE Task Committee on Geostatistical Techniques (1990a) 'Review of Geostatistics in Geohydrology I: Basic Concepts', *Journal of Hydraulic Engineering*, 116(5), pp. 612-632.

ASCE Task Committee on Geostatistical Techniques (1990b) 'Review of Geostatistics in Geohydrology II: Applications', *Journal of Hydraulic Engineering*, 116(5), pp. 633-658.

Atwood, D.F. and Gorelick, S.M. (1985) 'Hydraulic gradient control for groundwater contaminant control', *J. Hydrol.*, Amsterdam, 76, pp. 85–106.

Cameron, K. and Hunter, P. (2002) 'Using spatial models and kriging techniques to optimize long-term ground-water monitoring networks: a case study', *Environmetrics*, 13, pp. 629–656.

Carrera, J., Usunoff, E. and Szidarovsky, F. (1984) 'A method for optimal observation network design for groundwater management', *J. Hydrology*, 73, pp. 147–163.

Chang, L.C., Shoemaker, C.A. and Liu, P.L.F. (1992) 'Optimal time-varying pumping rates for groundwater remediation: Application of a constrained optimal control algorithm', *Water Resour. Res.*, 28(12), pp. 3157–3171.

Chang, L.C and Hsiao, C.T. (2002) 'Dynamic optimal ground water remediation including fixed and operation costs', *Ground Water*, 40(5), pp. 481-491.

Chan Hilton, A.B., and T.B. Culver (2000) 'Constraint handling for genetic algorithms in optimal remediation design', *J. Water Resour. Plan. Manage.*, 126(3), pp.128-137.

Cieniawski, S.E., Eheart, J.W. and Ranjithan, S. (1995) 'Using genetic algorithms to solve a multiobjective groundwater monitoring problem', *Water Resources Research*, 31(2), pp. 399-409.

Culver, T.B. and Shoemaker, C.A. (1992) 'Dynamic optimal control for groundwater remediation with flexible management periods', *Water Resour. Res.*, 28(3), pp. 629–641.

Culver, T.B. and Shoemaker, C.A. (1997) 'Dynamic optimal ground-water reclamation with treatment capital costs', *J. Water Resour. Plan. Manage.*, 123(1), pp. 23–29.

Datta, B. and Dhiman, S.D. (1996) 'Chance constrained optimal monitoring network design for pollutants in groundwater', *J. Water Res. Plng. And Mgmt, ASCE*, 122(3), pp. 180-188.

Davis, L., ed. (1991) *Handbook of Genetic Algorithms*, Nostrand Reinhold, New York, NY.

Dougherty, D.E. and Marryott, R.A. (1991) 'Optimal groundwater management 1. Simulated Annealing', *Water Resour. Res.*, 27(10), pp. 2493-2508

Goldberg, D.E. (1989) *Genetic Algorithms in Search, Optimization and Machine Learning*, Addison-Wesley, Reading, Mass..

Gorelick, S.M., Voss, C.I., Gill, P.E., Murray, W., Saunders, M.A. and Wright, M.H. (1984) 'Aquifer reclamation design: The use of contaminant transport simulation combined with nonlinear programming', *Water Resour. Res.*, 20(4), pp. 415–427.

Harrouni, E.K., Ouazar, D., Walters, G.A. and Cheng, A.H.D. (1996) 'Groundwater optimization and parameter estimation by genetic algorithm and dual reciprocity boundary element method', *Eng. Anal. Boundary Elem.*, 18(4), pp. 287–296.

Holland, J.H. (1975) *Adaption in Natural and Artificial Systems*, Univ. of Mich. Press, Ann Arbor.

Hsiao, C.T. and Chang, L.C. (2002) 'Dynamic Optimal Groundwater Management with Inclusion of Fixed Costs', *J. of Water Resources Planning and Management*, 128(1), pp. 57-65.

Huang, C. and Mayer, A.S. (1997) 'Pump-and-treat optimization using well locations and pumping rates as decision variables', *Water Res. Res.*, 33(5), pp. 1001-1012.

Jones, L., Willis, R. and Yeh, W.W.G. (1987) 'Optimal control of nonlinear groundwater hydraulics using differential dynamic programming', *Water Resour. Res.*, 23(11), pp. 2097-2106.

Karpouzos, D.K., Delay, F., Katsifarakis, K.L. and de Marsily, G. (2001) 'A multipopulation genetic algorithm to solve the inverse problem in hydrogeology', *Water Resour. Res.*, 37(9), pp. 2291-2302.

Kwanyuen B., and Fontane, D. (1998) 'Heuristic Branch-and-Bound Method for Ground Water Development Planning', *J. Water Resour. Plng. and Mgmt.*, 124(3), pp. 140-148.

Marryott R.A., Dougherty, D.E. and Stollar, R.L. (1993) 'Optimal groundwater management. 2. Application of simulated annealing to a field-scale contamination site', *Water Resour. Res.*, 29, pg. 847.

Maskey S., Jonoski, A. and Solomatine, D.P. (2002), *J. of Water Resources Planning and Management*, 128(6), pp. 431-440.

McKinney, D.C. and Lin, M.D. (1994) 'Genetic algorithm solution of groundwater management models', *Water Resour. Res.*, 30(6), pp. 1897–1906.

McKinney, D.C. and Lin, M.D. (1995) 'Approximate mixed-integer nonlinear programming methods for optimal aquifer remediation design', *Water Resour. Res.*, 31(3), pp. 731–740.

Meyer P.D., Valocchi, A.J. and Eheart, J.W. (1994) 'Monitoring network design to provide initial detection of groundwater contamination', *Water Res. Res.*, 30(9), pp. 2647-2659.

Minsker, B.S. and Shoemaker, C.A. (1998) 'Dynamic optimal control of in-situ bioremediation of ground water', *J. Resour. Plng. Mgmt., ASCE*, 124(3), pp. 149–161.

Morshed, J. and Kaluarachchi, J.J. (1998) 'Parameter estimation using artificial neural network and genetic algorithm for free-product migration and recovery', *Water Res. Res.*, 34(5), pp. 1101-1113.

Morshed, J. and Kaluarachchi, J.J. (2000) 'Enhancements to genetic algorithm for optimal ground-water management', *J. Hydrologic Eng.*, 5(1), pp. 67–73.

Reed, P., Minsker, B. and Valocchi, A.J. (2000) 'Cost-effective long-term groundwater monitoring design using a genetic algorithm and global mass interpolation', *Water Res. Res.*, 36(12), pp. 3731-3742.

Reed, P., Minsker, B. and Goldberg, D.E. (2000) 'Designing a competent simple genetic algorithm for search and optimization', *Water Res. Res.*, 36(12), pp. 3757-3762.

Reed, P. (2002) Striking the balance: Long-term groundwater monitoring design for multiple conflicting objectives, Ph.D. Thesis, University of Illinois.

Ritzel, B.J., Eheart, J.W. and Ranjithan, S. (1994) 'Using genetic algorithms to solve a multiple objective groundwater pollution containment problem', *Water Resour. Res.*, 30(5), pp. 1589-1603.

Rizzo, D.M. and Dougherty, D.E. (1996) 'Design optimization for multiple management period groundwater remediation', *Water Resour. Res.*, 32(8), pp. 2549-2561.

Rogers, L.L. and Dowla, F.U. (1994) 'Optimization of groundwater remediation using artificial neural networks with parallel solute transport modeling', *Water Resour. Res.,* 30, pp. 457-481.

Sen, M. and Stoffa, P.L. (1995) *Advances in exploration Geophysics: Global optimization methods in geophysical inversion*, Elsevier publication.

Smalley, J.B., Minsker, B.S. and Goldberg, D.E. (2000) 'Risk-based in situ bioremediation design using a noisy genetic algorithm', *Water Resour. Res.*, 36(10), pp. 3043-3052.

Solomatine, D.P. (1998) 'Genetic and other global optimization algorithms—comparison and use in calibration problems', *Proc. of 3rd Int. Conf. on Hydroinformatics, Balkema, Rotterdam,* The Netherlands, pp. 1021–1027.

Taghavi, S.A., Howitt, R.E. and Mariño, M.A. (1994) 'Optimal control of ground-water quality management: Nonlinear programming approach', *J. Water Resour. Plan. Manage.*, 120(6), pp. 962–982.

Takahashi, S. and Peralta, R.C. (1995) 'Optimal perennial yield planning for complex nonlinear aquifers: Methods and examples', *Adv. Water Resour.*, 18, pp. 49–62.

Wang, M. and Zheng, C. (1998) 'Application of genetic algorithms and simulated annealing in groundwater management: formulation and comparison', *Journal of American Water Resources Association*, 34(3), pp. 519-530.

Wagner, B. (1995) 'Recent advances in simulation-optimization groundwater management modeling', *Review of Geophysics*, 33 Supplement.

Yoon, J.H. and Shoemaker, C.A. (1999) 'Comparison of optimization methods for groundwater bioremedation', *J. Water Resour. Plan. Manage.*, 125(1), pp. 54–63.

Yoon, J.H. and Shoemaker, C.A. (2001) 'Improved Real-Coded GA for Groundwater Bioremediation', *J. Comp. in Civil Eng.*, 15, pg. 224.

Zheng, C. and Wang, P.P. (1996) 'Parameter structure identification using tabu search and simulated annealing', *Adv. Water Res.*, 19(4).

Zheng, C. and Wang, P.P. (1999) 'An integrated global and local optimization approach for remediation system design', *Water Resour. Res.*, 35(1), pp. 137–148.

Zheng, C. and Wang, P.P. (2002) 'A field demonstration of the simulation-optimization approach for remediation system design', *Ground Water*.

Zheng, C. and Wang, P.P. (2003) MGO: A modular groundwater optimize incorporating MODFLOW/MT3DMS, Documentation and User's Guide, May 2003.

In: New Developments in Sustainable Petroleum Engineering
Editor: Rafiq Islam
ISBN: 978-1-61324-159-2
© 2012 Nova Science Publishers, Inc.

Chapter 7

EFFECT OF OILFIELD SULPHATE SCALES ON THE PRODUCTIVITY INDEX

Fadairo A. S. Adesina[*], C. T. Ako, Omole Olusegun and Falode Olugbenga*

Covenant University, Ota, Nigeria, University of Ibadan, Nigeria

ABSTRACT

The precipitation and deposition of scale pose serious injectivity and productivity problems. Several models have been developed for predicting oilfield scales formation and their effect on deliverability of the reservoir to aid in planning appropriate injection water programme. In this study an analytical model has been developed for predicting productivity index of reservoir with incidence of scale deposition in the vicinity of the well bore.

NOMENCLATURES

B_o = Oil Formation volume factor, dimensionless
B_w = Water Formation volume factor, dimensionless
C = Concentration, g/m^3
h = Thickness, m
K = Instantaneous permeability, m^2
K_o = Initial permeability, m^2
K_{dep} = Deposition rate constant
r_w = Well bore Radius, m
R = Radial distance, m

---
[*] Corresponding e-mail: adesinafadairo@yahoo.com

S_w = Water saturation, dimensionless

S_{w_c} = Connate water saturation, dimensionless

T = Temperature, K

t = Production time, sec

μ_o = Oil Viscosity, cp

μ_w = Water Viscosity, cp

ϕ_o = Initial Porosity, dimensionless

ϕ = Instantaneous porosity, dimensionless

ρ = density, g/m³

INTRODUCTION

The magnitude of flow impairment induced by oilfield sulphates scale deposition around the well bore require description and classification of sulphate scales precipitation, scale build up (saturation), and their corresponding formation damage scenario at different operational and reservoir/brine parameters such as scale concentration in the brine, viscosity of brine, formation volume factor of the brine, solid scale density, injection rate, pressure drawdown, reservoir temperature, reservoir thickness, brine velocity and radial distance from the vicinity of the well bore (Fadairo et al, 2008; Civian, 2001). Major factors that influence the mixing and scale formation have been described in the literature (Fadairo, 2004; Oddo and Tomson, 1991).

The details of how sulphate scales are deposited in different locations in the reservoir near the well bore region and the consequent formation damage has also been presented (Fadairo, 2004, Fadairo et al, 2008).

Mixing of incompatible waters take place in the water contacted portion of the reservoir during water flooding (Fadairo, 2004; Fadairo et al, 2008; Civian, 2001; Bedrikovetsky et al., 2003a, 2004).

For instance, when formation water containing barium ion and seawater containing sulphate ions are mixed together, barium sulphate (precipitate) is formed in the swept zone inside the reservoir and is deposited as scale around the well bore during waterflooding.

In general the water-mixing zone moves from injection wells towards producers and precipitation of scales takes place in the mixing zone. Accumulation of precipitates does not depend upon mixing zone movement. The amount of scale precipitated at any point in the reservoir during such movement does not cause significant reduction in formation permeability (Civian, 2001; Bedrikovetsky et al., 2003a, 2004; Richards, 1968).

Diffusion and intensive mixing increase close to the production well due to increase in fluid velocity which also causes increase in the rate of chemical reaction in this region (Civian, 2001; Bedrikovetsky et al., 2003a, 2004).

These phenomenon lead to a higher rate of sulphate precipitation in the near well bore region than inside the reservoir. The overall effect of the above is that most of the sulphate precipitation and hence permeability reduction occur in the vicinity of the well (Fadairo et al, 2008; Bedrikovetsky et al., 2003b).

Studies (Fadairo et al, 2008; Rosario and Bezerra, 2001; Rocha, 2001) have shown that formation damage caused by sulphate precipitation can be better described as an exponential function of the depositional rate constant, salt concentration and production time rather than by a hyperbolic function of these variables as proposed by Bedrikovetsky et al[7]. Hence, there is the need to develop an expression for estimating the productivity index in wells producing from water-flooded reservoirs with possible incidence of sulphate scale formation based on the exponential function model. The purpose of this paper is to present the mathematical expression for achieving this purpose.

MODEL DEVELOPMENT

Consider the radial two-phase flow of oil and water at constant total flow rate q_t; saturated with solid-state particle at a location r, from the well bore. Assuming an idealized flow equation, the pressure gradient is expressed as follows:

$$\frac{dp}{dr} = \frac{q_t}{2\pi r h K_s} \left(\frac{k_{rw}}{\mu_w} + \frac{k_{ro}}{\mu_o} \right)^{-1} \quad (1)$$

Fadairo et al, 2008, recently expressed the effect of scale builds up on permeability variation based on exponential shape for both porosity and permeability damage function as

$$K_s = K_o [1 - \lambda_\phi S_s (1 - S_{w_i})]^{3.0} \quad (2)$$

where K_s is define as the instantaneous permeability as a result of solid scale saturation near the well bore region.

The derivation of equation (2) is expressed in appendix A.
Substituting equation (2) into (1) and re-arranging; we obtain

$$\frac{dp}{dr} = \frac{q_t}{2\pi r h K_o [1 - \lambda_\phi S_s (1 - S_{w_i})]^{3.0}} \left(\frac{k_{rw}}{\mu_w} + \frac{k_{ro}}{\mu_o} \right)^{-1} \quad (3)$$

$$dp = \frac{q_t}{2\pi r h K_o [1 - \lambda_\phi S_s (1 - S_{w_i})]^{3.0}} \left(\frac{k_{rw}}{\mu_w} + \frac{k_{ro}}{\mu_o} \right)^{-1} dr \quad (4)$$

Integrating equation (4); we have

$$\int_{p_i}^{p} dp = \frac{q_t}{2\pi h K_o [1 - \lambda_\phi S_s (1 - S_{w_i})]^{3.0}} \left(\frac{k_{rw}}{\mu_w} + \frac{k_{ro}}{\mu_o} \right)^{-1} \int_{r_w = r_w \exp(-2s)}^{R} \frac{1}{r} dr \quad (5)$$

$$\Delta p = \frac{q_t}{2\pi h K_o [1-\lambda_\phi S_s(1-S_{w_i})]^{3.0}} \left(\frac{k_{rw}}{\mu_w}+\frac{k_{ro}}{\mu_o}\right)^{-1} \ln\frac{R}{\bar{r}_w} \qquad (6)$$

Therefore

$$\frac{2\pi h K_o \Delta p}{q_t}\left(\frac{k_{rw}}{\mu_w}+\frac{k_{ro}}{\mu_o}\right)[1-\lambda_\phi S_s(1-S_{w_i})]^{3.0} = \ln\frac{R}{\bar{r}_w} \qquad (7)$$

The productivity index PI in a water-flooded reservoir with possible incidence of sulphate scale deposition around the well bore can be expressed as

$$PI \Rightarrow \frac{q_t}{\Delta p} = \frac{2\pi h K_o[1-\lambda_\phi S_s(1-S_{w_i})]^{3.0}}{\ln\frac{R}{\bar{r}_w}}\left(\frac{k_{rw}}{\mu_w}+\frac{k_{ro}}{\mu_o}\right) \qquad (8)$$

At initial production time, when t = 0 and $S_s = 0$ the expression degenerates to the normal Dupui formula for productivity index.

$$\frac{2\pi h K_o \Delta p^o}{q_t}\left(\frac{k_{rw}}{\mu_w}+\frac{k_{ro}}{\mu_o}\right) = \ln\frac{R}{\bar{r}_w} \qquad (9)$$

DISCUSSION OF RESULTS

The model was validated using the data of Haarberg *et al* (1991) on scale precipitation for given pore volume of sea water injected and fluid/ reservoir properties data[1,] (Fadairo, 2004; Fadairo et al, 2008; Richards, 1968) shown below in Tables1 and 2 respectively, as input into the model.

The result of the model show that productivity decline due to scale formation around the well bore is better described when the formation damage function was assumed to be an exponential function of the scale depositional rate constant, salt concentration and time than when hyperbolic shape was assumed as opined by Bedrikovetsky *et al* (2003a). The rate of decline of the productivity of a formation due to sulphate scale deposition may be sharp initially but usually the rate decreases exponentially as the pore volume of the seawater injected water increases (Civian et al 2001; Fadairo, 2004). The productivity decline caused by scale formation around the well bore was over estimated when the formation damage was assumed to be a hyperbolic function (figure 1). The higher the average scale precipitation in the mixing zone along the reservoir, the lower the well productivity index (Table 1).

Table 1. Amount of BaSO₄ and SrSO₄ precipitated as a function of pore volume of seawater injected (after Haarberg *et al* 1991)

Pore volume of seawater injected (%)	BaSO₄ Precipitate (g/m³)	SrSO₄ Precipitate (g/m³)
0	0.0	0.0
10	71.0	0.0
20	65.0	0.0
30	58.0	45.0
40	48.0	68.0
50	42.0	58.0
60	32.0	26.0
70	25.0	0.0
80	18.0	0.0
90	10.0	0.0
100	0.0	0.0

Table 2. Fluid and reservoir base case properties civan (2001) used as input in the scale prediction model

Pay thickness (h)	26m
Initial permeability	0.5922E-13m² (60mD)
Initial porosity	0.04
Reservoir pressure	36600kpa
Bottom hole pressure	22060kpa
Reservoir temperature	353K (80C)
Formation volume factor	0.254
Viscosity	0.0007Pa-s
Connate water saturation	0.2

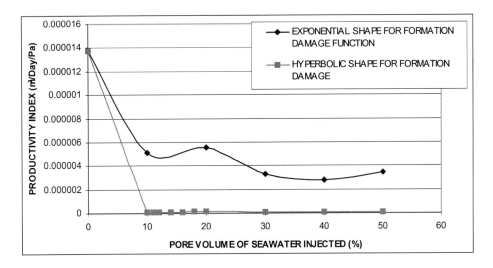

Figure 1. Productivity Index of a Well that Forms Sulphate Scale against Pore Volume of Seawater Injected.

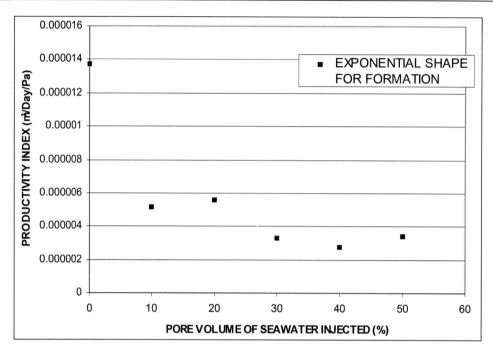

Figure 2. Productivity Index of a Well that Forms Sulphate Scales against Pore Volume of Seawater Injected.

Drastic productivity decline was observed at lower pore volume of seawater-injected ranges from 0% to 10% due to high permeability decline causes by deposited scale at low pore volume of water injected. This is in agreement with the studies reported in the literature ((Fadairo, 2004; Fadairo et al, 2008; Civian, 2001; Bedrikovetsky et al., 2003a, Tomson, 1991, Moghadasi et al., 2006).

This phenomenon might have been caused by heterogeneous nucleation that occurs at the early stage of scale build up around the well bore before re-dissolution of scale begins to take place at higher pore volume of injected water (Crabtee et al., 1999).

The increase in productivity index observed between 45% to 100% pore volume of water injected might have been due to scale re-dissolution (figure 2)

CONCLUSION

The following conclusions were drawn from the result of this study:

- At every given pore volume of sea water injected, the decline in productivity index for oil wells in water flooded reservoirs due to sulphate scale deposition depends upon oilfield solid scale saturation in the porous media.
- The rate of decline in productivity index due to sulphate scale deposition is the function of operational and reservoir/brine parameters such as scale concentration in the brine, viscosity of brine, formation volume factor of the brine, solid scale density, injection rate, pressure drawdown, reservoir temperature, reservoir thickness, brine velocity and radial distance from the well bore.

- The developed model is capable of predicting productivity index decline due to scale deposition around the well bore in water-flooded reservoirs.

APPENDIX A.
INSTANTANEOUS PERMEABILITY AS A RESULT OF SOLID SCALE SATURATION NEAR THE WELL BORE REGIO

Instantaneous local porosity can be defined as the difference between the initial porosity and damaged fraction of the pore spaces (Fadairo et al, 2008; Moghadasi et al., 2006).

That is $\phi_s = \phi_0 - \phi_d$ (A1)

Therefore

$$\phi_s = \phi_o - \frac{q^2 \left(\dfrac{dC}{dP}\right)_T . B . \mu . t . \lambda_k}{4\pi^2 r_s^2 h^2 K_o \rho} \quad (A2)$$

Damage fraction of the pore spaces ϕ_d can be defined as the ratio of the volume of scale deposited to bulk volume of the porous media or the fraction of minerals scale that occupied the total volume of porous media (Fadairo et al, 2008; Moghadasi et al., 2006).

That is $\phi_d = \dfrac{\text{volume of minerals scale deposited}}{\text{bulk volume of the porous media}}$

Also Fadairo (2004); recently expressed the fraction of mineral scale that occupied the pore spaces at different radial distance from the well bore as follows:

$$S_s = \frac{q^2 \left[\dfrac{dC}{dP}\right]_T . B_w . \mu_w . t . \lambda_k}{4\pi^2 . r_s^2 . h^2 . \phi_o \lambda_\phi K_o \rho (1 - S_{w_i})} \quad (A3)$$

Re-arranging equation (A3), we have:

$$\phi_o \lambda_\phi S_s (1 - S_{w_i}) = \frac{q^2 \left(\dfrac{dC}{dP}\right)_T . B_w . \mu_w . t . \lambda_k}{4\pi^2 . r_s^2 . h^2 K_o \rho} \quad (A4)$$

where λ_ϕ and λ_k can be defined as porosity damage coefficient and permeability damage coefficient respectively.

That is $\lambda_\phi = \exp(-K_{dep}.C.t)$ and $\lambda_k = \exp(3K_{dep}.C.t)$ (A5)

Substituting equation (A3) in equation (A2), we have

$$\phi = \phi_o - \phi_o \lambda_\phi S_s (1 - S_{w_i})$$ (A6)

Dividing both side of equation (A6) by ϕ_o, we have

$$\frac{\phi_s}{\phi_o} = 1 - \lambda_\phi S_s (1 - S_{w_i})$$ (A7)

Considering the relationship between the initial permeability and instantaneous permeability as a function of altered porosity and initial porosity defined by Frank et al (1991) as:

$$\frac{K_s}{K_o} = \left(\frac{\phi}{\phi_o}\right)^3$$ (A8)

Instantaneous permeability can be expressed as Fadairo, (2004):

$$K_s = K_o [1 - \lambda_\phi S_s (1 - S_{w_i})]^{3.0}$$ (A9)

Equation (A9) expresses the effect of scale build up on permeability variation at different operational parameters and reservoir/brine parameters such as scale concentration in the brine, viscosity of brine, formation volume factor of the brine, solid scale density, injection rate, pressure drawdown, reservoir temperature, reservoir thickness, connate water saturation against injection time and radial distance from the well bore vicinity.

APPENDIX B.
PIRSON'S CORRELATION

Average relative permeability was expressed as a function of average water saturation was given below as follows (Ahmed, 1989):

$$S_w^* = \frac{S_w - S_{w_c}}{1 - S_{w_c}}$$ (B-1)

$$K_{r_w} = \sqrt{S_w^*}\, S_w^3 \qquad \text{(B-2)}$$

$$K_{r_o} = (1-S_w^*)\left[1-\left(S_w^*\right)^{0.25}\sqrt{S_w}\right]^{0.5} \qquad \text{(B-3)}$$

REFERENCES

Ahmed, T. (1989) *Introduction to Petroleum Reservoir Analysis*, Gulf Publishing Company. Houston TX.

Atkinson, G., Raju, K. and Howell, R.D. (1991) "*The Thermodynamics of Scale Prediction*", SPE 21021, pp.209–215, presented at the Society of Petroleum Engineers international Symposium on Oilfield chemistry Anaheim, Canada.

Bedrikovetsky, P.G., Gladstone, P.M., Lope, Jr., Rosario, F.F., Silva, M.F., Bezerra, M.C. and Lima, E.A. (2003) *Oilfield Scaling Part I: Mathematical and Laboratory Modeling*, SPE 81127, presented in SPE American and Caribbean Petroleum Engineering Conference in Port of Spain Trinidad West Indies.

Bedrikovetsky, P.G., Gladstone, P.M., Lope, Jr., Rosario, F.F., Silva, M.F., Bezerra, M.C. and Lima, E.A. (2003b) *Oilfield scaling part II: Productivity Index Theory*, SPE 81128, SPE Latin American and Caribbean Petroleum Engineering Conference, Port-of-Spain, Trinidad and Tobago.

Bedrikovetsky, P.G., Gladstone, P.M., Lope, Jr., Rosario, F.F., Silva, M.F., Bezerra, M.C. and Lima, E.A. (2004) *Barium Sulphate Oilfield Scaling: Mathematical and Laboratory Modeling*, SPE 87457, SPE International Symposium on Oilfield Scale, Aberdeen, United Kingdom.

Civan, F. (2001) *Modeling Well Performance under Non Equilibrium Deposition Condition*, SPE 67234, presented at SPE production and Operations Symposium Oklahoma, USA.

Crabtee, M., Eshinger, D., Fletches, P., Johnson, A. and King, G (1999) Fighting Scale – Removal and Prevention, Oilfield Review, Schlumberger Autumn.

Fadairo, A.S.A. (2004) Prediction Scale Build Up Rate around the Well Bore, MSc Thesis, Department of Petroleum Engineering, University of Ibadan, Nigeria.

Fadairo, A.S.A. Omole, O. and Falode O. (2008) 'Effect of Oilfield Scale on Mobility Ratio', SPE 114488 Gas Technology Symposium 2008 Joint Conference held in Calgary, Alberta, Canada, 16–19 June 2008.

Frank, F., Chang and Civan, F. (1991) *Modeling of Formation Damage Due to Physical and Chemical Interaction between Fluid and Reservoir Rock*, SPE 22856. SPE Annual Technical Conference and Exhibition, Dallas, Texasn

Haarberg, T., Selm, I., Granbakken, D.B., Østvold, T., Read, P. and Schmidt, T. (1992) Scale Formation in Reservoir and Production Equipment during Oil Recovery II" equilibrium Model, *SPE Journal Production Engineering*, Volume 7, Number 1n

Moghadasi, J., Sharif, A., Kalantari, A.M. and Motaie, E. (2006a) *A New Model to Describe Particle Movement and Deposition in Porous Media*, SPE 99391, presented at 15th SPE Europe Conference and Exhibition, Vienna, Austria.

Oddo, J.E. and Tomson, M.B. (1994) 'Why scale forms in the oilfield and method of predict it', SPE 21710, *SPE Journal Production and Facilities*, Volume 9, Number 1n

Richards, S.G. (1968) Effect of Brine Concentration and Pressure Drop Gypsum Scaling in Oil Well, SPE 1830, *Journal of Petroleum Technology*, Volume 20, Number 6n

Robert, B.E (1997) Effect of Sulphur Deposition on Gas Well Performance, SPE 36707, *SPE Annual Technical Conference and Exhibition in Denver*, Colorado.

Rocha, A.A., Frydman, M., Fontoura, S.A.B., Rosario, F.F. and Bezzarra, M.C. (2001) *Numerical Modelling of Salt Precipitation During Produced Water Re-injection, SPE 68336*, International Symposium on Oilfield Scale, Aberdeen, United Kingdomn

Rosario, F.F. and Bezerra (2001) *Scale Potential of a Deep Water Field Water Characterization and Scaling Assessment,* SPE 68332, International Symposium on Oilfield Scale, Aberdeen, United Kingdom.

Chapter 8

A MONTE-CARLO SIMULATION BASED ASSESSMENT APPROACH FOR ANALYZING ENVIRONMENTAL HEALTH RISKS FROM B.T.E.-CONTAMINATED GROUNDWATER

L. Liu[1]*, J. B. Li[2] and S. Y. Cheng[3]

[1]Department of Civil and Resource Engineering, Dalhousie University, Halifax, Canada
[2]Environmental Science and Engineering Program, University of Northern British Columbia, Prince George, BC, Canada
[3]College of Environmental and Energy Engineering, Beijing University of Technology, Beijing, China

ABSTRACT

In North America, numerous aquifers as fresh drinking water supply sources have been contaminated from various sources such as septic systems, leaking underground storage tanks, spills or improper disposal of industrial chemicals, and leachate from solid and hazardous waste landfills. A major task associated with the contamination of aquifers is to develop effective tools to assess and determine the health risks to individuals potentially consuming the water from these aquifers. This poses many obstacles due to the complexity of the environment that the contaminant is spilled in and the population consuming the water. This paper presents a Monte Carlo simulation based methodology which is capable of considering many variations present in the natural environment and human populations. The methodology is tested using a hypothetical aquifer and population case. Downgradient contaminant concentrations resulting from a leaking underground storage tank, containing petroleum, are calculated using an analytical solution to the two-dimensional advection-dispersion transport model. Contaminants under consideration include benzene, toluene, and ethylbenzene (BTE) which can cause deleterious health effects. Parameters and variables in the model are considered as

---

* Correspondence Author: Dr. L. Liu, Department of Civil and Resource Engineering, Dalhousie University, 1360 Barrington St. Halifax, NS B3J 1Z1 Canada. Email: Lei.Liu@Dal.Ca

random numbers. Contaminant ingestion dose is calculated using the stochastic exposure-dose model. Random population variables are used to give a distribution for contaminant ingestion dose for each contaminant. The chronic non-cancer hazard index is used to determine the risk associated with the ingestion of non-carcinogenic contaminants. The cancer risk is calculated using the slope factor, given by the USEPA, for the carcinogenic contaminants. Results of the case study indicate that environmental health risks can be effectively analyzed through the developed methodology. They are useful for supporting the related risk-management and remediation decisions.

Keywords: assessment, groundwater contamination, health risk, Monte Carlo simulation, uncertainty.

1. INTRODUCTION

Contamination of drinking water supplies, resulting from environmental pollution, poses a serious threat to many North Americans. Countless numbers of aquifers across North America have been contaminated from point sources such as septic systems, leaking underground storage tanks, spills or improper disposal of industrial chemicals, and leachate from solid and hazardous waste landfills (Kidd, 2002). 198 million North Americans rely on groundwater as their sole source of drinking water (Kidd, 2002) with 439,385 confirmed contaminant releases from underground storage tanks in the United States alone (USEPA, 2003). Petroleum products are the most common substances stored in underground storage tanks and thus merit attention. Many of the compounds found in petroleum pose a threat to public health and safety when released into aquifers supplying drinking water. As public understanding of groundwater contamination has grown, so has concern about the quality of drinking water from groundwater sources. Thus to protect and inform the public in cases where this valuable resource has been contaminated it is essential to have the necessary tools to determine the human ingestion rates and risk associated with ingestion in addition to contaminant transport and fate.

Many researchers have studied the transport and fate of contaminants in groundwater. Consequently, various methods are available to professionals attempting to mathematically determine down-gradient contaminant concentrations. Methodologies to reach a solution vary from one-dimensional analyses to more complex multi-dimensional analyses with various models allowing for naturally occurring variations within the aquifer. Numerical and analytical solutions are both used to evaluate contaminant transport in groundwater. For example, Arial and Liao (1996) provide various solutions to the advection-dispersion equation defined for a steady-state two-dimensional velocity field with first-order decay function and time dependent dispersion coefficient. Vogel et al. (2000) present a methodology to model flow and transport in a two-dimensional dual-permeability system with spatially variable hydraulic properties. Sheu and Chen (2002) present a study with the key goal of attaining a solution to the contaminant transport problem in a scheme that provides precision and stability. Solutions are achieved by applying the semi-discretization method to approximate time and spatial derivatives separately. The finite element solution was compared with the analytical solution to help prove its validity with strikingly similar results. In addition, Kolmogorov-Dmitriev model provides a methodology other than the advection-dispersion model for analyzing groundwater transport. Ferrara et al (1999) provide a comparison between the Kolmogorov-Dmitriev model and advection-dispersion model. The Kolmogorov-Dmitriev model provides a simple way to study the evolution of contaminant in time and space. Work completed by Ferrara *et al* (1999) demonstrates that contaminant movement in groundwater under non-equilibrium parameters is more realistically described by the Kolmogorov-Dmitriev model than the advection-dispersion model.

It has long been known that the contamination of drinking water supplies is a cause for environmental and health concerns. Assessing human exposure to contaminated drinking water poses difficulties made worse by a number of factors, including lack of measured exposure data, uncertainty about the initiation of contamination, duration of exposure, direction and movement of contaminant plume, lack of tools and procedures involving multidisciplinary approaches needed to identify the link between environmental studies and increased health risk to humans (Masila et al, 1997). Exposure assessment can be approached directly or indirectly. Direct approaches include personal exposure monitoring and biological markers of exposure. Personal exposure monitoring environmental contaminants refers to the collection of samples at the boundary between the exposure medium and the human receptor (MacIntosh and Spengler, 2000). Indirect approaches include environmental sampling, combined with exposure factor information, modeling and questionnaires (MacIntosh and Spengler, 2000).

The Agency for Toxic Substances and Disease Registry (ATSDR) of the U.S. Department of Human Health Services determines contaminant exposure waste sites using an exposure investigation approach to assess past, present, and potential future human exposure (Masila *et al*, 1999). The exposure investigation approach used by ATSDR combines indirect and direct approaches through implementation of three methods to obtain information about contaminant ingestion. Biomedical testing involving the testing of blood and urine samples provides information on recent and current exposure to contaminants. Environmental testing through the sampling and analysis of contaminated soil, water, and/or air focused on areas where people may be exposed to the contaminants. Lastly, exposure-dose reconstruction analyses using environmental sampling information and numerical models to reconstruct historical contaminant levels individuals may have been exposed to and the use of the models to predict future exposure. Information obtained through various measures can be input into numerical models to answer questions about duration and levels of contamination of drinking water and adverse affects to individual health. With human exposure examined, it is then extremely important to have the ability to evaluate the risk associated with exposure to a contaminant. Risk associated with the ingestion of contaminated groundwater poses many problems due to uncertainties. Uncertainties exist in the models used to assess risk, physical and chemical properties of the site, exposure conditions at the site, relative toxicity, and other properties.

In this study, examination of contaminant transport is completed through the application of the analytical form of the two-dimensional advection-dispersion equation proposed by Huyakorn *et al.* (1987). The equation considers aquifer and contaminant properties to calculate downgradient contaminant concentrations. Transport and fate of gasoline components (B.T.E.: benzene, toluene, and ethybenzene) with the potential to inflict negative health effects on humans over extended exposure periods, was considered. The transport and fate simulation is examined by a hypothetical case with a leaking underground storage tank. In doing so a methodology has been laid out that allows for the complete examination of a contaminated aquifer under varying conditions. A complete analysis would involve considering all of the chemical components of gasoline posing potential negative health effects resulting from exposure. Contaminant ingestion is considered using the stochastic exposure-dose model. This model provides the daily contaminant ingestion rate for an individual by considering various exposure parameters. This daily contaminant ingestion rate is then used to calculate the risk of developing negative health effects through the application of the chronic non-cancer hazard index and the linear low dose cancer risk equation. Natural variations in soil properties, chemical properties, and human populations provide difficulty in calculating risk, which were presented as random distributions. During the overall procedure, the Monte Carlo simulation approach is employed to deal with random parameters and to present the range of risk associated with contaminant exposure for a given population.

2. METHODOLOGY

2.1. Contaminant Transport

An examination of benzene, toluene, and ethylbenzene transport in an aquifer, which is assumed homogenous and isotropic, was completed by using a two-dimensional advection-dispersion equation proposed by Huyakorn *et al.*(1987). The equation can be described as follows:

$$v_s \frac{\partial C}{\partial x} - D_{xx} \frac{\partial^2 C}{\partial x^2} - D_{yy} \frac{\partial^2 C}{\partial y^2} + \lambda RC + \frac{I_R RC}{b} = 0 \tag{1}$$

This partial differential equation utilizes the following boundary conditions:

$$C(0, y) = C_0 \exp\left(-\frac{y^2}{2\sigma^2}\right) \tag{2}$$

$$C(\infty, y) = C(x, \infty) = C(x, -\infty) = 0 \tag{3}$$

The following variables can be used to non-dimensionalize equations (1) and (2):

$$X = \frac{vx}{D_{xx}} \tag{4}$$

$$Y = \frac{y}{\sigma} \tag{5}$$

$$D = \frac{D_{xx} D_{yy}}{\sigma^2 v^2} \tag{6}$$

$$\varphi = \frac{RD_{xx}}{v^2}\left(\lambda + \frac{I_R}{b}\right) \tag{7}$$

$$c = \frac{C}{C_0} \tag{8}$$

Equations (1) and (2) now become:

$$\frac{\partial c}{\partial X} - \frac{\partial^2 c}{\partial X^2} - \frac{\partial^2 c}{\partial Y^2} + \varphi c = 0 \tag{9}$$

$$c(0,y) = \exp\left(-\frac{Y^2}{2}\right) \tag{10}$$

The analytical form of the two-dimensional steady state advection-dispersion equation can then be given as follows:

$$c(X,Y) = \frac{\exp\left\{\dfrac{\dfrac{X}{2}\left[1-(1+4\varphi)^{1/2}\right]-Y}{2+\dfrac{4DX}{(1+4\varphi)^{1/2}}}\right\}}{\left[1+\dfrac{2DX}{(1+4\varphi)^{1/2}}\right]^{1/2}} \tag{11}$$

where C = aqueous-phase solute concentration (mg/L); x = longitudinal coordinate (m); y = transverse coordinate (m); v_s = groundwater seepage velocity (m/d); D_{xx} = hydrodynamic dispersion in x-direction; D_{yy} = hydrodynamic dispersion in y-direction, (m²/d) $D_{yy} = 0.01 D_{xx}$ (Gelhar et al., 2002); R = retardation factor; λ = first-order biodegradation rate constant (d⁻¹); I_R = net regional infiltration rate (m/d); b = saturated zone thickness (m); C_0 = maximum concentration at plume source (mg/L); σ = standard deviation in source width, (m), $\sigma = \sqrt{2Dt_t}$ (Fetter, 2001).

Seepage velocity is calculated as follows:

$$v_s = -Ki/\eta_e \tag{12}$$

where K = hydraulic conductivity; i = hydraulic gradient, $i = dH/dL = (h_2 - h_1)d$; h_1 = upgradeint head; h_2 = downgradient head; d = distance between pumping wells; η_e = effective porosity.

Hydrodynamic dispersion is the sum of molecular diffusion and mechanical dispersivity and takes into account mechanical mixing and diffusion. Typical values range from 10^{-7}-10^{-2} m²/d (Alshawabkeh, 2003). For the purpose of calculations used in this study, a smaller normally distributed range of 10^{-2}-10^{-4} m²/d will be used. Calculation of hydrodynamic dispersion is represented by the following equation (Fetter, 2001):

$$D_{xx} = \alpha_x v_s + D^* \tag{13}$$

where α_x = longitudinal dispersivity; D^* = dynamic dispersivity.

First-order biodegradation constants for BTE was obtained by the Buscheck and Alcantar method. It employs a one-dimensional analytical solution to interpret the steady-state configuration of the contaminant plume along the flow path using the following equation (Corseuil et al., 2002):

$$\lambda = \left(\frac{v_c}{4\alpha_x}\right)\left\{\left[1 + 2\alpha_x\left(\frac{j}{v_s}\right)\right] - 1\right\} \tag{14}$$

where v_c = retarded contaminant velocity in flow path direction = v_s/R; α_x = longitudinal dispersivity; j/v_s = slope of the log-linear plot of contaminant concentration vs. distance downgradient along the flow path.

The retardation factor is described as follows (Weaver, 2002):

$$R = 1 + \frac{\rho_b k_d}{\eta_e} \tag{15}$$

where R = retardation factor; ρ_b = bulk density = $\rho_s(1-\eta_e)$; ρ_s = solids density; η_e = effective porosity; k_d = (soil) distribution coefficient = $f_{oc} k_{oc}$; f_{oc} = fraction organic carbon; k_{oc} = organic carbon/water partition coefficient.

Infiltration rate can be determined with many different models. The Explicit Green-Ampt Model is the first physically based equation describing infiltration of water into soil and used in this study (Salvucci and Entekhabi, 1994). It relies on the following assumptions to be applicable: 1) water content profile is of a piston-type with well-defined wetting front, 2) antecedent water content distribution is uniform and constant, 3) water content drops abruptly to its antecedent value at the front, 4) soil-water pressure head at wetting front is h_f, 5) soil water pressure head at the surface, h_s, is equal to depth of the ponded water, and 6) soil in the wetted region has constant properties (Williams et al., 1988).

The following equations describe the Explicit Green-Ampt Model (Salvucci and Entekhabi, 1994):

$$I = \left[\begin{array}{l}\left(1 - \frac{\sqrt{2}}{3}\right)t + \frac{\sqrt{2}}{3}\sqrt{\chi t + t^2} + \left(\frac{\sqrt{2}-1}{3}\right)\chi[\ln(t+\chi) - \ln\chi] \\ + \frac{\sqrt{2}}{3}\chi\left[\ln\left(t + \frac{\chi}{2} + \sqrt{\chi t + t^2}\right) - \ln\left(\frac{\chi}{2}\right)\right]\end{array}\right] K \tag{16}$$

$$I_R = \left(\frac{\sqrt{2}}{2}\zeta^{-1/2} + \frac{2}{3} - \frac{\sqrt{2}}{6}\zeta^{1/2} + \frac{1-\sqrt{2}}{3}\zeta\right)K \tag{17}$$

with

$$\chi = \frac{(h_s - h_f)(\theta_s - \theta_o)}{K} \tag{18}$$

and

$$\zeta = \frac{t}{t + \chi} \tag{19}$$

where I = cumulative infiltration at time t (cm); I_R = net regional infiltration rate (cm/h); K = hydraulic conductivity (cm/h); t = time (hr); h_s = ponding depth or capillary pressure head at the surface (cm); h_f = capillary pressure head at the wetting front (cm); θ_s = saturated volumetric water content (cm^3/cm^3); θ_o = initial volumetric water content (cm^3/cm^3).

h_f is calculated by the following equation (Brakensiek and Onstad, 1977):

$$h_f = \frac{(2+3\omega)}{(2+3\omega)-1} h_e \tag{20}$$

where $h_e = \frac{1}{2} h_b$; h_e = air exit head (cm); ω = constant exponent of the Brooks-Corey water retention model (= 1.68) (Williams, 1998).

2.2. Contaminant Ingestion Dose

When the contaminant oral daily intake via exposure pathways has been estimated, the dose can be determined. For chemicals, the dose to humans is measured as milligrams per kilogram per day (mg/kg-day). In this case, the "kilogram" refers to the body weight of an adult individual. When a chemical dose is calculated, the length of time an individual is exposed to a certain concentration is important.

To assess off-site doses, it is assumed that the exposure duration occurs over 30 years. Such exposures are called "chronic" in contrast to short-term exposures, which are called "acute" (USOERR, 1990). The stochastic exposure-dose model (USOERR, 1990) provides a method for calculating chronic daily intake as follows:

$$CDI = \frac{C \times IR \times FG \times EF \times ED}{BW \times AT} \tag{21}$$

where CDI = chronic daily ingestion dose of contaminant (mg/kg-day); C = concentration of contaminant in groundwater (mg/L); IR = intake rate groundwater (L/day); FG = ingestion fraction groundwater; EF = exposure frequency (days/year); ED = exposure duration (years); BW = body weight (kg); AT = averaging exposure time (days) (70 years for carcinogen, $365 \times ED$ for chronic health risk).

2.3. Analysis of Risk

Risks for benzene, toluene, and ethylbenzene oral exposure were calculated using the procedure laid out by the U.S. Office of Emergence and Remedial Responses (USOERR, 1989). The chronic non-cancer hazard index is calculated by:

$$HI = CDI/RfD \tag{22}$$

where HI = hazard index; RfD = chronic reference dose.

Cancer risk is calculated as follows by using the linear low-dose cancer risk equation:

$$CR = CDI \times SF \tag{23}$$

where CR = cancer risk; SF = slope factor.

3. RESULTS

3.1. Solute Aqueous Phase Down-gradient Concentration

Concentrations at the well were calculated for benzene, toluene, and ethylbenzene using the advection dispersion equation (12). Calculations were completed in Excel© with a Monte Carlo simulation done using Crystal Ball©. For each simulation 2,000 trials were completed giving the concentration distribution. Negative value results were excluded. In the Monte Carlo simulation hydrodynamic dispersion in the x-direction was considered randomly distributed between 10^{-2}-10^{-4} m^2/d with a mean of 5×10^{-3} m^2/d. Typical values range from 10^{-7}-10^{-2} m^2/d (Alshawabkeh, 2003). Hydrodynamic dispersion in the y-direction was approximated to be 0.001 times hydrodymanic dispersion in the x-direction (Gelhar et al., 1992). The maximum concentration at the plume source was also considered a random variable in the advection-dispersion equation. The ranges of values for each contaminant were determined by considering the composition of gasoline and are summarized in Table 1. Concentrations were based on the distribution of benzene, toluene, and ethylbenzene compounds in gasoline (Table 1) as described by Christansen and Elson (1996), where benzene, toluene, and ethylbenzene typically makes up 8.6% of gasoline by weight with gasoline having a solubility of 300 mg/L. The first-order biodegradation constant, calculated using the Buscheck and Alcantar method, was used for the calculation of the concentration distribution (Table 1). The forecasted concentration distribution for each compound was based on the assumptions that the aquifer was homogenous with the properties of a silty loam soil texture. Properties of a silty loam soil texture were determined through a literature review (Carsel and Parish, 1988; Rawls et al, 1989; MPCA, 2000; Corseuil et al., 2002). Retardation factor was calculated using the described soil properties using the method described by Weaver (2002) in equation (16). The calculated retardation factors for each contaminant are summarized in Table 1. The seepage velocity, of the hypothetical aquifer, was calculated to be 1.34 m/day using equation (12). The net regional infiltration rate was calculated to be

0.212 m/day based on the aquifer properties using equations (16) through (19). A summary of the aquifer properties and calculated values is provided in Table 2.

Table 1. B.T.E. concentrations calculated by the advection-dispersion equation

Value		Benzene	Toluene	Ethylbenzene
Retardation factor, R		14.24	176.41	48.84
Fraction of gasoline		0.02	0.05	0.02
First-order biodegradation rate constant, λ, (year-1)		1.75	2.63	1.72
Maximum concentration, C_0	(mg/L)			
	Max	2	5	2
	Mean	5	35	5
	Min	10	65	10

Table 2. Calculated and assumed aquifer properties

Property	PDF	Value
Porosity, η, (cm^3/cm^3)	Deterministic	0.501
Effective porosity, η_e, (cm^3/cm^3)	Deterministic	0.486
Pore size index, ε	Deterministic	0.41
Residual water content, (cm^3/cm^3)	Deterministic	0.015
Arithmetic bubbling pressure, h_b, (cm)	Deterministic	50.87
Hydraulic conductivity, K, (cm/hr)	Deterministic	0.68
Saturated volumetric water content, Θ_s	Deterministic	0.45
Initial volumetric water content, Θ_o	Deterministic	0.067
Air entry head, h_h, (cm)	Deterministic	50
Bulk Density, ρ_b, (g/cm^3)	Deterministic	1.55
Fraction of Organic Carbon, f_{oc}	Deterministic	0.05
Exponent of the Brooks-Corey water retention model, ω	Deterministic	1.68
Air exit head, h_e, (cm)	Deterministic	-25.435
Capillary pressure head at wetting front, h_f, (cm)	Deterministic	-29.646
Net regional infiltration rate, IR, (m/day)	Deterministic	0.213
Standard deviation in source width, σ, (m)	Deterministic	0.863
Transverse coordinate, y, (m)	Deterministic	5
Travel distance in x-dir, x, (m)	Deterministic	75
Saturated zone thickness, b, (m)	Deterministic	75
Groundwater seepage velocity, v_s, (m/day)	Deterministic	1.34
Hydrodynamic dispersion in x-dir, D_{xx}, (m^2/day)	Normal	.01-.0001
Hydrodynamic dispersion in y-dir, D_{yy}, (m^2/day)	Normal	$.001 \times D_{xx}$

The distribution of concentrations of benzene, toluene, and ethylbenzene in the groundwater at the well are presented in Figures 1, 2, and 3 below. The results give the range of concentrations for each compound expected at the well in addition to the probability and frequency of each concentration occurring.

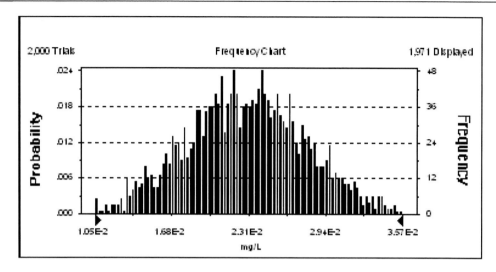

Figure 1. Aqueous phase concentration (mg/L) distribution for benzene at the well.

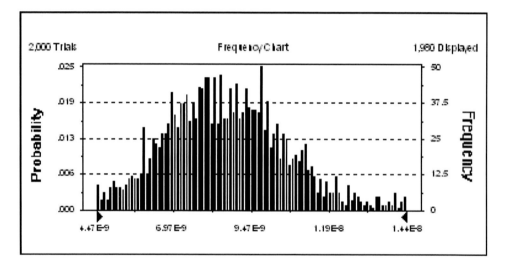

Figure 2. Aqueous phase concentration (mg/L) distribution for toluene at the well.

In each figure the number of trials is listed as well as the number of trials displayed. The number of trials displayed indicates that some results have been excluded due to negative values resulting from the calculations. The resulting probability distribution for benzene is given in Figure 1 with a mean of 2.30×10^{-2} mg/L, a range minimum of 8.53×10^{-3} mg/L and a range maximum of 3.57×10^{-2} mg/L. The resulting probability distribution for toluene is given in Figure 2 with a mean of 1.49×10^{-3} mg/L, a range minimum of 5.99×10^{-4} mg/L, and a range maximum of 2.72×10^{-3} mg/L.

The resulting probability distribution for ethylbenzene is given in Figure 3 with a mean of 8.69×10^{-9} mg/L, a range minimum of 3.17×10^{-9} mg/L, and a range maximum of 1.79×10^{-8} mg/L.

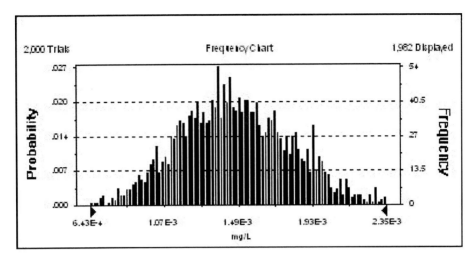

Figure 3. Aqueous phase concentration (mg/L) distribution for ethylbenzene at the well.

3.2. Stochastic Exposure-Dose Model

The stochastic exposure dose model (Equation (21)) was used to determine the chronic daily ingestion dose of contaminant. A Monte Carlo simulation was completed using Crystal Ball© with a number of random variables.

Table 3. Stochastic exposure dose model constraints and values

Constants	Value
Intake rate water (L/day)	2
Exposure frequency (days/year)	365
Body weight (kg)	70
Averaging exposure time (day)	
Carcinogen (70 years)	25550
Chronic health risk (ED365)	variable
Random Variables	
Ingestion fraction groundwater	
min	0.1
mean	0.5
max	0.9
Exposure duration (years)	
min	1
mean	25
max	70

The resulting aqueous phase concentration distribution for each contaminant was entered into the equation as a random variable. Other random variables in the equation include ingestion fraction and exposure duration. Ingestion fraction represents the percentage of daily fluid intake coming from the contaminated well. Exposure duration is the time that the individual is exposed to the contaminant. The exposure duration is assumed to be randomly distributed between 1 year and 70 years, representing an entire lifetime, with a mean of 25 years. A summary of the values used for each of the contaminants is given in Table 3.

The resulting probability distribution for benzene is given in Figure 4 with a mean of 1.17×10^{-4} mg/kg-day, a range minimum of 3.81×10^{-5} mg/kg-day, and a range maximum of 2.25×10^{-4} mg/kg-day.

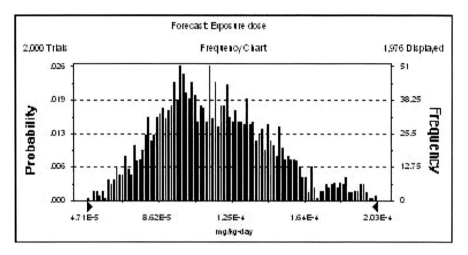

Figure 4. Forecasted exposure dose (mg/kg-day) distribution for benzene.

The resulting probability distribution for toluene is given in Figure 5 with a mean of 1.25×10^{-10} mg/kg-day, a range minimum of 4.09×10^{-11}, and a range maximum of 2.65×10^{-10} mg/kg-day.

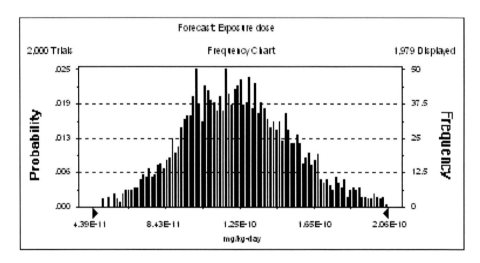

Figure 5. Forecasted exposure dose (mg/kg-day) distribution for toluene.

The resulting probability distribution for ethylbenzene is given in Figure 6 with a mean of 2.12×10^{-5} mg/kg-day, a range minimum of 5.66×10^{-6} mg/kg-day, and a range maximum of 4.67×10^{-5} mg/kg-day.

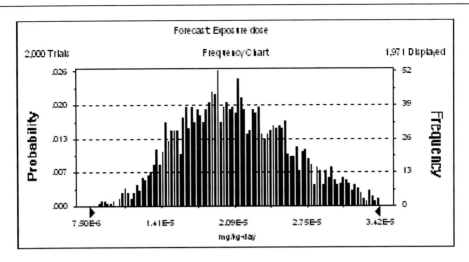

Figure 6. Forecasted exposure dose (mg/kg-day) distribution for ethylbenzene.

3.3. Analysis of Risk

The risk associated with exposure to benzene, toluene, and ethylbenzene was calculated using the form provided in Equations (22) and (23) (USORER, 1989). The linear low-dose cancer risk equation was used to calculate the cancer risk associated with the ingestion of benzene in water. Again Crystal Ball© was used to complete a Monte Carlo simulation, with the calculated exposure dose distribution as a random variable, to give a frequency distribution of the associated risk. The slope factor used, as determined by the USEPA (2004), was a random variable ranging from 1.5×10^{-2} to 5.5×10^{-2} per (mg/kg)/day with a mean of 3.5×10^{-2} per (mg/kg)/day. The resulting cancer risk frequency distribution is given in Figure 7 with a mean probability of 4.1×10^{-6} or 1 in 244,000, a range minimum 1.43×10^{-6} or 1 in 6.99×10^{5}, and a range maximum 9.52×10^{-6} or 1 in 1.05×10^{5}.

The maximum and minimum risk associated with benzene ingestion was also determined using the extremes of the random variables applied to the entire model. For the extreme maximum exposure case the risk of being affected by cancer associated with the ingestion of benzene was calculated to be 1 in 4283 individuals (Table 4).

For the extreme minimum exposure case, the risk was calculated to be 1 in 8.71×10^{18} (Table 4) individuals being affected by cancer because of benzene ingestion. The chronic non-cancer hazard index (Equation (22)) was used to calculate the health risk associated with ingestion of toluene and ethylbenzene. The calculated risk associated with equation (22) is considered unacceptable with the chronic non-cancer index equals or exceeds a value of 1.0. Individual risk resulting from ingestion of each contaminant was calculated separately and in combination to give the risk distribution. Monte Carlo simulation in Crystal Ball© was used to complete this calculation.

Chronic reference dose used for toluene and ethylbenzene was 2×10^{-1} mg/kg/day and 1×10^{-1} mg/kg/day respectively. The resulting forecasted risk distribution for toluene is shown in Figure 8 with a mean of 2.48×10^{-11}, a range minimum 7.58×10^{-12}, and a range maximum 4.90×10^{-11} all much less than 1.

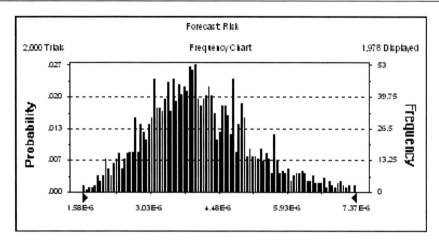

Figure 7. Forecasted per person cancer risk distribution for benzene ingestión.

Table 4. Minimum and maximum risk values associated with extreme low and extreme high benzene ingestion exposure

Random Variable	Minimum	Maximum
Hydrodynamic dispersion in x-direction, Dxx, (m^2/d)	0.0001	0.01
Maximum concentration at plume source, C_0, (mg/l)	2	10
Aqueous-phase solute concentration, C, (mg/L)	1.88×10^{-13}	1.50×10^{-1}
Ingestion fraction groundwater	0.1	0.90
Exposure duration (years)	1	70
Exposure dose, (mg/kg-day)	7.65×10^{-18}	4.24×10^{-3}
Slope factor ((mg/kg)/day)	0.015	0.055
Risk (individual)	1.15×10^{-19}	2.33×10^{-4}
1/Risk (1 in X number of people will get cancer)	8.71×10^8	4.28×10^3

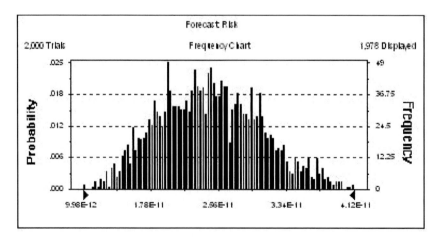

Figure 8. Forecasted chronic non-cancer health hazard ingestion risk distribution associated with exposure to toluene.

The resulting forecasted risk distribution for ethylbenzene is shown in Figure 9 with a mean of 2.13×10^{-6}, a range minimum 8.73×10^{-7}, and a range maximum 4.84×10^{-6} all much less than 1.

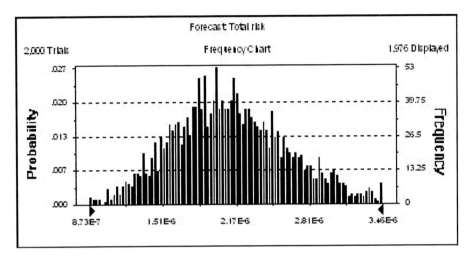

Figure 9. Forecasted chronic non-cancer health hazard ingestion risk distribution associated with exposure to ethylbenzene.

The resulting forecasted combined risk distribution for toluene and ethylbenzene is shown in Figure 10 with a mean of 2.13×10^{-6}, a range minimum 8.73×10^{-7}, and a range maximum 4.84×10^{-6} all much less than 1. These values are the same as for ethylbenzene due to the small values associated with toluene risk.

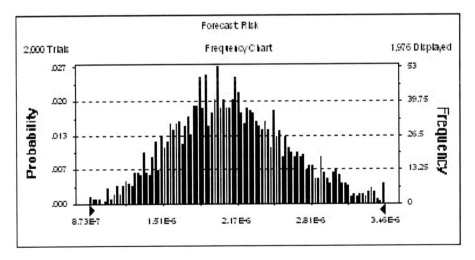

Figure 10. Forecasted chronic non-cancer health hazard ingestion risk distribution associated with exposure to toluene and ethylbenzene.

The maximum and minimum risk of chronic health problems resulting from ingestion of toluene and ethylbenzene were calculated using the maximum and minimum values for the random variables. The results are summarized in Table 5.

Table 5. Minimum and maximum risk values associated with extreme low and extreme high benzene ingestion exposure

Random Variable	Ethylbenzene Minimum	Toluene Minimum	Ethylbenzene Maximum	Toluene Maximum
Hydrodynamic dispersion in x-direction, D_{xx}, (m²/d)	0.0001	0.0001	0.01	0.01
Maximum concentration at plume source, C_0 (mg/L)	2	5	10	65
Aqueous-phase solute concentration, C, (mg/L)	1.21×10^{-14}	2.42×10^{-20}	9.76×10^{-3}	5.54×10^{-8}
Ingestion fraction groundwater	0.1	0.1	0.9	0.9
Exposure duration, (years)	1	1	70	70
Exposure dose, (mg/kg-day)	3.47×10^{-17}	6.93×10^{-23}	2.51×10^{-4}	1.42×10^{-9}
Risk, (individual)	3.47×10^{-18}	1.39×10^{-23}	2.51×10^{-5}	2.85×10^{-10}
Combined Risk	\multicolumn{2}{c}{3.47×10^{-18}}	\multicolumn{2}{c}{2.51×10^{-4}}

4. DISCUSSIONS

4.1. Discussions on B.T.E. Concentration

Determining the health risk associated with the ingestion of contaminants transported in groundwater proves difficult when accuracy is concerned. Many factors from aquifer properties, contaminant properties, and population factors make determining the actual health risk with definite accuracy near impossible. This study considers a hypothetical site where an underground storage tank leaking gasoline into an aquifer. Gasoline contains numerous contaminants that pose health hazards if ingested. For completing this risk assessment this study only examined benzene, toluene, and ethylbenzene. A methodology has been laid out for determining the distribution of risk associated with the ingestion of these contaminants. A complete health risk assessment would involve the examination of each of the individual contaminants that posed a health hazard. The methodology presented in this paper is applicable to various conditions where the calculation and presentation of the risk associated with a leaking contaminant.

Contaminant transport was analyzed using the two-dimensional advection dispersion presented by Huyakorn et al. (1987). For the purpose of this study, hydrodynamic dispersion in the x-direction and source contaminant concentration were considered random variables. Hydrodynamic dispersion in the y-direction was assumed to be 0.001 multiplied by hydrodynamic dispersion in the x-direction as discussed in the methodology section of the paper. All other factors were considered constant. This assumption was made to simplify the problem used for completing this study. Real site conditions would have more variability. Time is not a factor in this equation as it assumes a steady state condition in the aquifer because sufficient time has elapsed to make this assumption plausible.

The two-dimensional advection dispersion equation requires the initial source concentration of the contaminant. This requires the examination of petroleum properties.

Brandes (1999) gives a gasoline solubility range from 100mg/L to 300 mg/L. Thus, the solubility of individual constituents would be less than these values. An initial petroleum concentration of 300mg/L was used to determine benzene, toluene, and ethylbenzene concentrations at the source. The aqueous phase concentration distributions given by Crystal Ball© can be seen in Figures 1, 2, and 3. The most notable concentration distribution is that of toluene which is negligible with a mean value of 8.69×10^{-9} mg/L, while the calculated means for benzene and ethylbenzene are 2.03×10^{-2} mg/L and 1.49×10^{-3} mg/L respectively. The reason for this dramatic difference in concentrations is a result of the difference in the organic carbon water partition coefficients and decay rates. The organic carbon water partition coefficient is used to calculate the retardation factor, which significantly affects the concentration down gradient. The calculated retardation factor for toluene is 176.41 as compared with 14.24 and 48.24 for benzene and ethylbenzene respectively (Table 1). The half-life of toluene also plays a significant role in the lower concentration values at the well. In the Buscheck and Alcantar method, toluene's half-life approaches half the value of benzene and ethylbenzene. The combination of these factors results in the extremely low aqueous phase concentration distribution for ethylbenzene downgradient at the well. Low concentrations associated with ethylbenzene suggests that the contaminant will have little effect on human health.

Benzene is the only contaminant that exceeds the Canadian Drinking Guidelines for safe drinking water making the water unsafe to drink (FPTCDW, 2003). The maximum possible concentration of benzene is 1.50×10^{-1} mg/L far above the provided guideline (the standard of Benzene = 5.00×10^{-4} mg/L). Benzene concentrations at the well are above the guideline for the entire range of calculated values as seen in Figure 1 with a mean of 2.30×10^{-2} mg/L, a range minimum of 8.53×10^{-3} mg/L and a range maximum of 3.57×10^{-2} mg/L. As a result, this water would be considered unsafe to drink when considering the drinking water guidelines. The guidelines presented for toluene (2.4×10^{-3} mg/L) and ethylbenzene (2.4×10^{-2} mg/L) are only aesthetic objectives. Toluene concentrations are far below the guideline value even at the maximum concentration possible (5.54×10^{-8} mg/L) (Table 5). Ethylbenzene does not exceed the maximum allowable guideline concentration of 2.40×10^{-2} mg/L when considering the maximum possible concentration of 9.76×10^{-3} mg/L (Table 5). Thus, the health risk associated with the ingestion of water containing the lower contaminant concentrations is very low.

4.2. Discussions on Ingestion Dosage

Chronic daily ingestion dose as calculated by the stochastic exposure dose model gives a widely distributed range of values for all three contaminants. This is a result of the random variables applied to the stochastic exposure dose model. Exposure duration for a hypothetical population was considered normally distributed between 1 and 70 years with a mean value of 25 years. To determine exposure duration for a real situation a survey of the population would be necessary. Ingestion fraction is also considered a random variable with a normal distribution between 10 percent and 90 percent with a mean of 75 percent. In a real situation, all of the values input into the stochastic exposure-dose model would be considered randomly distributed except for averaging exposure time for a carcinogen, which is given by the USEPA, although, the value, 70 years, provided by the USEPA is a conservative value with the average North American life span approaching 80 years. Because individual lives are so variable, the results for a distribution of chronic daily dosages are expected to be widely ranging.

The benzene predicted probability distribution for chronic daily ingestion dosage is given in Figure 4 with a mean of 1.17×10^{-4} mg/kg-day, a range minimum of 3.81×10^{-5} mg/kg-day, and a range maximum of 2.25×10^{-4} mg/kg-day. The maximum allowable daily ingestion dose given by the USEPA (2004) is 4.0×10^{-3} mg/kg-day. This would seem to suggest that ingestion of benzene with the predicted ingestion dosage range would be acceptable. The maximum possible daily dose of 4.24×10^{-3} mg/kg-day must also be considered. This value exceeds the USEPA's maximum allowable daily ingestion dose and thus the chances of negative health effects would be increased.

The toluene predicted probability distribution for chronic daily ingestion dosage is given in Figure 5 with a mean of 1.25×10^{-10} mg/kg-day, a range minimum of 4.09×10^{-11} mg/kg-day, and a range maximum of 2.65×10^{-10} mg/kg-day. This range represents extremely small values especially when considering the possible negative health affects associated with toluene was combined with the risk associated with ethylbenzene. This is due to the substances having similar non-carcinogenic health effects. Ethylbenzene's predicted probability distribution for chronic daily ingestion dosage is given in Figure 6 with a mean of 2.12×10^{-5} mg/kg-day, a range minimum of 5.66×10^{-6} mg/kg-day, and a range maximum of 4.67×10^{-5} mg/kg-day representing significantly larger values than toluene.

4.3. Discussions on Health Risks

Both cancer and non-cancer risks were assessed. Toluene and ethylbenzene are non-carcinogens but can have deleterious effects on the liver and kidneys if ingested in high enough quantities over long enough periods. The hypothetical case gives a combined chronic non–cancer hazard index probability distribution for toluene and ethylbenzene with a mean of 2.13×10^{-6}, a range minimum 8.73×10^{-7}, and a range maximum 4.84×10^{-6} as shown in Figure 10. A value of 1 or greater is required to cause health concerns to be raised. Even when considering the maximum possible chronic non-cancer hazard index of 2.51×10^{-4} there is negligible risk associated with the ingestion of toluene and ethylbenzene found in the drinking water. It is important to note that toluene has little to no-effect on the chronic non-cancer hazard index as downgradient concentrations are so small due to a very large organic water carbon partition coefficient. The results are as expected given the properties of the contaminants and the regulations regarding them.

The risk associated with the ingestion of benzene in the concentrations found at the well is unacceptable. The USEPA states the risk of developing cancer should be less than 1 in 1,000,000. The forecast frequency distribution for cancer risk is given in Figure 7 with a mean probability of 4.1×10^{-6} or 1 in 244,000, a range minimum 1.43×10^{-6} or 1 in 6.99×10^{5}, and a range maximum 9.52×10^{-6} or 1 in 1.05×10^{5}. For the extreme maximum exposure case the risk of being affected by cancer associated with the ingestion of benzene was calculated to be 1 in 4283 individuals (Table 4). All of these values exceed the maximum risk for development of cancer, even at the low end of the distribution. Benzene does pose a much greater threat to human health than toluene or ethylbenzene, therefore, the water in this case should not be consumed, even in small quantities. Concentrations of benzene are much higher downgradient than the toluene or ethylbenzene, resulting in the increased risk of developing health problems. Benzene concentrations are higher because benzene has much smaller organic water carbon partition coefficient (83 mL/g) than toluene (1100 mL/g) or ethylbenzene (300 mL/g).

CONCLUSION

This study provides a valuable insight into the determination of the risk related to contaminant ingestion as well as the reflection of extensive uncertainties associated with the risk assessment process through a Monte-Carlo simulation approach. The methodology presented is capable of considering many of the variations that are present in the natural environment and human populations. It is tested using a hypothetical aquifer and population case. Downgradient contaminant concentrations resulting from a leaking underground storage tank, containing petroleum, are calculated using an analytical solution to the two-dimensional advection-dispersion transport model. Chemicals under consideration include benzene, toluene, and ethylbenzene which can cause deleterious health effects. Parameters and variables in the model are considered as random numbers. Contaminant ingestion dose is calculated using the stochastic exposure-dose model. Random population variables are used to give a distribution for contaminant ingestion dose for each contaminant. The chronic non-cancer hazard index is used to determine the risk, associated with the ingestion of contaminants that are not carcinogenic. The cancer risk is calculated using the slope factor, determined by the USEPA, for the carcinogenic contaminants.

Improvements could be made through the provision of data from a real site. This would allow the calculation of downgradient contaminant concentrations using known aquifer and contaminant properties for a site already studied. Additionally, using this methodology to determine downgradient contaminant concentrations, contaminant ingestion dose, and health risk with real data would allow for comparison of results from the use of different methodologies. Being able to compare results would help prove reliability of the methodology used in this study. It could also be used to determine weakness and limitations associated with one methodology as compared with another. It would be interesting to use a population with know statistics as well. Having defined distributions for variables associated with the stochastic exposure-dose model would give results that are more meaningful by allowing one to see how a real population would be affected by a contaminant spill.

ACKNOWLEDGMENT

The authors would like to thank the Natural Science and Engineering Research Council of Canada for funding this research. This study was also supported by the Natural Science Foundation of China (50609008), the Natural Science Foundation of Beijing (8082022), and the Major State Basic Research Development Program of China (2005CB724200 and 2005CB724201). The authors are grateful to the editors and the anonymous reviewers for their insightful comments.

REFERENCES

Alshawabkeh, A.N. 2003. Contaminant fate and transport. Research and industrial collaboration conference, November 18-19, [online]. Available: http://www.censsis.neu.edu/RICC/2003 /WorkshopOne.html [March 3, 2004].

Aral, M.M., Liao, B. 1996. Analytical solutions for tow-dimensional transport equation with time-dependent dispersion coefficients. *Journal of Hydrologic Engineering*. vol.1, no.1, pp.20-32.

Brakensiek, D.L., Onstad, C.A. 1977. Parameter estimation of the Green and Ampt infiltration equation. *Water Resource Research*. vol.13, pp.1009-1012.

Brandes, R.J. 1999. A review of groundwater modeling for selected aquifers underlying the longhorn partners pipeline, [online]. Available: www.epa.gov-earth1r6-6en-xp-lppapp.pdf [March 20, 2004].

Carsel, R.F., Parrish, R.S. 1988. Developing joint probability distributions of soil water retention characteristics. *Water Resource Research*. vol.24, pp.755-769.

Christansen, J.S., Elson, J. 1996. Soil and groundwater pollution from BTEX, [online]. Available: http://www.cee.vt.edu/program_areas/environmental/teach/gwprimer/btex/btex.html#1.%20Introduction [March 20, 2004].

Corseuil. H.X., Schneider, M., Rosario, M. 2002. Natural attenuation rates of ethanol and BTEX compounds in groundwater contaminated with gasohol. *Appropriate Environmental and Solid Waste Management and Technologies for Developing Countries*. vol.4, pp.2121-2128.

Ferrara, A., Marseguerra, M., Zio, E. 1999. A comparison between the advection-dispersion and the kollmogrov-dimitriev model for groundwater contaminant transport. *Annals of Nuclear Energy*. vol.26, no.12 ,pp.1083-1096.

Fetter, C.W. 2001. *Applied Hydrogeology 4th Edition*. Prentice Hall, Englewood Cliffs, NJ, 598 pp.

FPTCDW (Federal-Provincial-Territorial Committee on Drinking Water). 2003. Summary of guidelines for Canadian drinking water quality. Health Canada.

Gelhar, L.W., Welty, C., Rehfeldt, K.R. 1992. A critical review of data in field-scale dispersion in aquifers. *Water Resource Research*. vol.28, no.7, pp.1955-1974.

Huyakorn, P.S., Ungs, M.J., Mulkey, L.A. and Sudicky, E.A. 1987. A three dimensional analytical method for predicting leachate migration. *Groundwater*. vol.25, no.5, pp.588-598.

Kidd, J. 2002. Groundwater: A North American Resource. Expert workshop on freshwater in North America, January 4, 2002, [online]. Available: http://www.cec.org/files/pdf/LAWPOLICY /water_disucssion-e1.pdf [April 1, 2004]

MacIntosh, D.L., Spengler, J.D. 2000. International programme on chemical safety: human exposure assessment. United Nations Environment Programme, International Labor Organization, World Health Organization. WHO Library Cataloguing ISBN 92 4 157214 0, [online]. Available: http://www.inchem.org/documents/ehc/ehc/ehc214.htm#Section Number:3.5 [March 5, 2004].

Masila, M.L., Aral, M.M., Williams, R.C. 1997. Exposure assessment using analytical and numerical models: a case study. Practice Periodical of Hazardous, Toxic, and Radioactive *Waste Management*. vol.1, no.2, pp.50-60.

MPCA (Minnesota Pollution Control Agency). 2000. Leaking petroleum storage tanks, assessment of natural attenuation at petroleum release sites. *Fact sheet* 3.21, April 2000.

Rawls, W.J., Ahuja L. R., Brakensiek D. L. 1989. Estimation of soil hydraulic properties from soils data. *Proceedings of the international workshop on indirect methods for estimating the hydraulic properties of unsaturated soils.* October 11-13, 1989.

Sheu, T.W.H., Chen, Y.W. 2002. Finite element analysis of contaminant transport in groundwater. *Applied Mathematics and Computations.* vol.127, no.23-24. pp.23-43.

Salvucci, G.D., Entekhabi, D. 1994. Explicit expressions for Green-Ampt (Delta function diffusivity) infiltration rate and cumulative storage. *Water Resource Research.* vol.30, pp.2661-2663.

USEPA (U.S. Environmental Protection Agency). 2003. UST Program Facts, [online]. Available: http://www.epa.gov/swerust1/pubs/ustfacts.pdf [2004, March 25].

USEPA (U.S. Environmental Protection Agency). 2004. IRIS (Integrated Risk Information System), [online]. Available: http://www.epa.gov/iris/ [2004, March 25].

USOERR (U.S. Office of Emergency and Remedial Responses). 1990. Guidance on Remedial Actions for Superfund Sites with PCB Contamination. Rep. No. EPA/540/G-09/007, U.S. Office of Emergency and Remedial Responses, Washington, D.C.

USOERR (U.S. Office of Emergency and Remedial Responses). 1989. *Risk Assessment Guidance for Superfund* Volume 1 Human Health Evaluation Manual (Part A). Rep. No. EPA/540/1-89/002, U.S. Office of Emergency and Remedial Responses, Washington, D.C.

Vogel, T. Gerke, T.T, Zhang, R., Genuchten, V. 2000. Modeling flow and transport in a two-dimensional dual-permeability system with spatially variable hydraulic properties. *Journal of Hydrology.* vol.238, no.1-2, pp.78-89.

Weaver, J. 2002. Modeling Subsurface Petroleum Hydrocarbon Transport, United States Environmental Protection Agency (USEPA), [Online]. Available: http://www.epa.gov/athens/learn2model/part-two/onsite/retard.htm [2004, March 6].

Williams, J.R., Ouyang, Y., Chen, J., Ravi, V. 1998. Estimation of Infiltration Rate in the Vadose Zone: *Application of Selected Mathematical Models* Volume II. USEPA (United States Environmental Protection Agency), February 1998.

In: New Developments in Sustainable Petroleum Engineering
Editor: Rafiq Islam

ISBN: 978-1-61324-159-2
© 2012 Nova Science Publishers, Inc.

Chapter 9

EXACT SOLUTIONS OF SYSTEMS OF LINEAR INTEGRO-DIFFERENTIAL EQUATIONS USING THE HPM

Hossein Aminikhah[1,] and Jafar Biazar[2]*

[1]Department of Mathematics, School of Mathematical Sciences,
Shahrood University of Technology, Shahrood, Iran
[2]Department of Mathematics, Faculty of Sciences,
University of Guilan, Rasht, Iran

ABSTRACT

In this paper, we introduce a new version of homotopy perturbation method to obtain exact solutions of systems of linear integro-differential equations. Theoretical considerations are discussed. Some examples are presented, to illustrating the efficiently and simplicity of the method.

Keywords: new homotopy perturbation method, systems of integro-differential equations

1. INTRODUCTION

Integro-differential equation has attracted much attention and solving this equation has been one of the interesting tasks for mathematicians. These equations has been found to describe various kind of phenomena such as wind ripple in the desert , nono-hydrodynamics, drop wise consideration and glass-forming process [1-4].

The homotopy perturbation method is a powerful device for solving functional equations. The method has been used by many authors to handle a wide variety of scientific and engineering applications to solve various functional equations. In this method the solution is

---

[*]Corresponding author: E-mail address: hossein.aminikhah@gmail.com, aminikhah@shahroodut.ac.ir (H. Aminikhah), biazar@guilan.ac.ir (J. Biazar).

considered as the summation of an infinite series which converges rapidly to the accurate solutions. Considerable research works have been conducted recently in applying this method to a class of linear and non-linear equations. This method was further developed and improved by He and applied to nonlinear oscillators with discontinuities [5], nonlinear wave equations [6], boundary value problems [7], limit cycle and bifurcation of nonlinear problems [8], and many other subjects [9-12]. It can be say that He's homotopy perturbation method is a universal one, is able to solve various kinds of nonlinear functional equations. For examples it was applied to nonlinear Schrödinger equations [13], to nonlinear equations arising in heat transfer [14], to the quadratic Ricatti differential equation [15], and to other equations [16-22]. Authors of [23, 24] employed He's homotopy perturbation method to compute an approximation to the solution of system of Volterra integral equations and non-linear Fredholm integral equation of second kind.

In this article a new homotopy perturbation method is introduced to obtain exact solutions of systems of integro-differential equations. To demonstrate this method, some examples are given.

2. New Version of Homotopy Perturbation Method for Systems of Integro-Differential Equations

A system of integro-differential can be considered in general as the following

$$\frac{d\mathbf{x}(t)}{dt} = \mathbf{F}(t, \mathbf{x}(t)) + \int_{t_0}^{t} \mathbf{K}(s, t, \mathbf{x}(s))\, ds, \quad t_0 \geq 0,$$
$$\mathbf{x}(t_0) = \mathbf{x}_0, \tag{1}$$

where

$$\mathbf{x}(t) = (x_1(t), x_2(t), \ldots x_n(t))^T,$$
$$\mathbf{K}(s, t, \mathbf{x}(t)) = (k_1(s, t, \mathbf{x}(s)), k_2(s, t, \mathbf{x}(s)), \ldots, k_n(s, t, \mathbf{x}(s)))^T.$$

If $\mathbf{K}(s, t, \mathbf{x}(t))$ and $\mathbf{F}(t, \mathbf{x}(t))$ be linear, the system (1) can be represented as the following simple form

$$\frac{dx_i}{dt} = f_i(t) + \sum_{j=1}^{n}\left(w_{i,j}(t)\, x_j(t) + \int_{t_0}^{t} k_{i,j}(s, t)\, x_j(s)\, ds\right), \quad x_i(t_0) = \alpha_i, \ i = 1, 2, \ldots, n. \tag{2}$$

For solving system (2), by new homotopy perturbation method, we construct the following homotopy

$$(1-p)\left(\frac{dX_i}{dt} - x_{i,0}\right) + p\left(\frac{dX_i}{dt} - f_i(t) - \sum_{j=1}^{n}\left(w_{i,j}(t)\, x_j(t) + \int_{t_0}^{t} k_{i,j}(s, t)\, x_j(s)\, ds\right)\right) = 0, \tag{3}$$

or equivalently,

$$\frac{dX_i}{dt} = x_{i,0} - p\left(x_{i,0} - f_i(t) - \sum_{j=1}^{n}\left(w_{i,j}(t)x_j(t) + \int_{t_0}^{t} k_{i,j}(s,t)x_j(s)\,ds\right)\right). \quad (4)$$

Applying the inverse operator, $L^{-1} = \int_{t_0}^{t}(\cdot)\,dt$ to both sides of equation (4), we obtain

$$\begin{aligned}X_i(t) &= \alpha_i + \int_{t_0}^{t} x_{i,0}(s)\,ds \\ &- p\left(\int_{t_0}^{t} x_{i,0}(s)\,ds - \int_{t_0}^{t} f_i(s)\,ds - \sum_{j=1}^{n}\left(\int_{t_0}^{t} w_{i,j}(s)x_j(s)\,ds + \int_{t_0}^{t}\int_{t_0}^{\tau} k_{i,j}(s,\tau)x_j(s)\,ds\,d\tau\right)\right).\end{aligned} \quad (5)$$

Suppose the solutions of system (5) have the following form

$$X_i(t) = X_{i,0}(t) + pX_{i,1}(t) + p^2 X_{i,2}(t) + \cdots, \quad i = 1,2,\ldots,n. \quad (6)$$

where $X_{i,j}(t)$, $i = 1,2,\ldots,n$ and $j = 0,1,2,\ldots$ are functions which should be determined.

Now suppose that the initial approximations to the solutions $X_{i,0}(t)$ or $x_{i,0}(t)$ have the form

$$X_{i,0}(t) = x_{i,0}(t) = \sum_{j=0}^{\infty} \alpha_{i,j} P_j(t), \quad i = 1,2,\ldots,n, \quad (7)$$

where $\alpha_{i,j}$ are unknown coefficients and $P_0(x), P_1(x), P_2(x),\ldots$ are specific functions.

Substituting (6) into (5) and equating the coefficients of p with the same power leads to

$$\begin{aligned}p^0 &: X_{i,0}(t) = \alpha_i + \sum_{j=0}^{\infty} \alpha_{i,j} \int_{t_0}^{t} P_j(s)\,ds, \\ p^1 &: X_{i,1}(t) = -\sum_{j=0}^{\infty} \alpha_{i,j} \int_{t_0}^{t} P_j(s)\,ds + \int_{t_0}^{t} f_i(s)\,ds \\ &\quad + \sum_{j=1}^{n}\left(\int_{t_0}^{t} w_{i,j}(s) X_{j,0}(s)\,ds + \int_{t_0}^{t}\int_{t_0}^{\tau} k_{i,j}(s,\tau) X_{j,0}(s)\,ds\,d\tau\right), \\ p^2 &: X_{i,2}(t) = \sum_{j=1}^{n}\left(\int_{t_0}^{t} w_{i,j}(s) X_{j,1}(s)\,ds + \int_{t_0}^{t}\int_{t_0}^{\tau} k_{i,j}(s,\tau) X_{j,1}(s)\,ds\,d\tau\right), \\ &\vdots \\ p^j &: X_{i,j-1}(t) = \sum_{j=1}^{n}\left(\int_{t_0}^{t} w_{i,j}(s) X_{j,j-1}(s)\,ds + \int_{t_0}^{t}\int_{t_0}^{\tau} k_{i,j}(s,\tau) X_{j,j-1}(s)\,ds\,d\tau\right), \\ &\vdots\end{aligned} \quad (8)$$

Now if these equations be solved in a way that $X_{i,1}(t) = 0$, then equations (8) result in $X_{i,2}(t) = X_{i,2}(t) = \cdots = 0$, therefore the exact solution can be obtained by using

$$x_i(t) = X_{i,0}(t) = \alpha_i + \sum_{j=0}^{\infty} \alpha_{i,j} \int_{t_0}^{t} P_j(s) \, ds. \tag{9}$$

It is worthwhile to note that if $f_i(t)$ and $x_{i,0}(t)$ are analytic at $t = t_0$, then their Taylor series

$$x_{i,0}(t) = \sum_{n=0}^{\infty} a_n (t - t_0)^n, \, f_i(t) = \sum_{n=0}^{\infty} a_n^* (t - t_0)^n, \tag{10}$$

can be used in Eqs. (8), where $a_0^*, a_1^*, a_2^*, \ldots$ are known coefficients and $a_0, a_1, a_2, \ldots$ are unknown ones, which must be computed.

We would explain this method by considering several examples.

3. EXAMPLES

In this section we present two examples. These examples are considered to illustrate the NHPM for systems of integro-differential equations.

Example 1. Consider the following system of integro-differential equations with the exact solutions $x_1(t) = e^t$ and $x_2(t) = e^{-t}$,

$$\begin{aligned}
\frac{dx_1(t)}{dt} &= t^4 - t^3 - 2t^2 - 6 + (3t^2 - 6t + 7) x_1(t) + 2t^2(t+1) x_2(t) \\
&\quad + \int_0^t \left((s^3 - t^3) x_1(s) + t^2(s^2 - t^2) x_2(s) \right) ds, \quad x_1(0) = 1, \\
\frac{dx_2(t)}{dt} &= -t^4 - 3t^2 + 2 + 2(t-1) x_1(t) + (2t^4 + 2t^3 + 2t^2 - 1) x_2(t) \\
&\quad + \int_0^t \left((s^2 - t^2) x_1(s) + t^2(s^2 + t^2) x_2(s) \right) ds, \quad x_2(0) = 1.
\end{aligned} \tag{11}$$

For solving system (11), by NHPM, we construct the following homotopy

$$\begin{aligned}
\frac{dX_1(t)}{dt} &= x_{1,0}(t) - p(x_{1,0}(t) - t^4 + t^3 + 2t^2 + 6 - (3t^2 - 6t + 7) X_1(t) - 2t^2(t+1) X_2(t) \\
&\quad - \int_0^t \left((s^3 - t^3) X_1(s) + t^2(s^2 - t^2) X_2(s) \right) ds), \\
\frac{dX_2(t)}{dt} &= x_{2,0}(t) - p(x_{2,0}(t) + t^4 + 3t^2 - 2 - 2(t-1) X_1(t) - (2t^4 + 2t^3 + 2t^2 - 1) X_2(t) \\
&\quad - \int_0^t \left((s^2 - t^2) X_1(s) + t^2(s^2 + t^2) X_2(s) \right) ds.
\end{aligned} \tag{12}$$

Assuming that,

$$x_{1,0}(t) = \sum_{n=0}^{\infty} \alpha_n P_n(t), x_{2,0}(t) = \sum_{n=0}^{\infty} \beta_n P_n(t), P_i(t) = t^i, X_1(0) = X_2(0) = 1.$$

By integration of equation (12) we have

$$X_1(t) = 1 + \sum_{n=0}^{\infty} \frac{\alpha_n}{n+1} t^{n+1} - p\big(\sum_{n=0}^{\infty} \frac{\alpha_n}{n+1} t^{n+1} - \frac{t^5}{5} + \frac{t^4}{4} + \frac{2t^3}{3} + 6t$$
$$- \int_0^t \big((3s^2 - 6s + 7) X_1(s) + 2s^2(s+1) X_2(s)\big) ds$$
$$- \int_0^t \int_0^\tau \big((s^3 - \tau^3) X_1(\tau) + \tau^2(s^2 - \tau^2) X_2(\tau)\big) ds\, d\tau\big), \qquad (13)$$

$$X_2(t) = 1 + \sum_{n=0}^{\infty} \frac{\beta_n}{n+1} t^{n+1} - p\big(\sum_{n=0}^{\infty} \frac{\beta_n}{n+1} t^{n+1} + \frac{t^5}{5} + t^3 - 2t$$
$$- \int_0^t \big(2(s-1) X_1(s) + (2s^4 + 2s^3 + 2s^2 - 1) X_2(s)\big) ds$$
$$- \int_0^t \int_0^\tau \big((s^2 - \tau^2) X_1(\tau) + \tau^2(s^2 + \tau^2) X_2(\tau)\big) ds\, d\tau\big).$$

Suppose the solutions of system (13) have the following form

$$X_i(t) = X_{i,0}(t) + p X_{i,1}(t) + p^2 X_{i,2}(t) + \cdots, \quad i = 1, 2, \qquad (14)$$

where $X_{i,j}(t)$, $i = 1, 2$ and $j = 0, 1, 2, \ldots$ are functions which should be determined.

Substituting (14) into (13) and equating the coefficients of p with the same powers leads to

$$p^0 : \begin{cases} X_{1,0}(t) = 1 + \sum_{n=0}^{\infty} \frac{\alpha_n}{n+1} t^{n+1}, \\ X_{2,0}(t) = 1 + \sum_{n=0}^{\infty} \frac{\beta_n}{n+1} t^{n+1}, \end{cases}$$

$$p^1 : \begin{cases} X_{1,1}(t) = -\sum_{n=0}^{\infty} \frac{\alpha_n}{n+1} t^{n+1} + \frac{t^5}{5} - \frac{t^4}{4} - \frac{2t^3}{3} - 6t + \int_0^t \big((3s^2 - 6s + 7) X_{1,0}(s) + 2s^2(s+1) X_{2,0}(s)\big) ds \\ \quad + \int_0^t \int_0^\tau \big((s^3 - \tau^3) X_{1,0}(\tau) + \tau^2(s^2 - \tau^2) X_{2,0}(\tau)\big) ds\, d\tau, \\ X_{2,1}(t) = -\sum_{n=0}^{\infty} \frac{\beta_n}{n+1} t^{n+1} - \frac{t^5}{5} - t^3 + 2t + \int_0^t \big(2(s-1) X_{1,0}(s) + (2s^4 + 2s^3 + 2s^2 - 1) X_{2,0}(s)\big) ds \\ \quad + \int_0^t \int_0^\tau \big((s^2 - \tau^2) X_{1,0}(\tau) + \tau^2(s^2 + \tau^2) X_{2,0}(\tau)\big) ds\, d\tau, \end{cases}$$

$$p^m : \begin{cases} X_{1,m}(t) = \int_0^t \big((3s^2 - 6s + 7) X_{1,m-1}(s) + 2s^2(s+1) X_{2,m-1}(s)\big) ds \\ \quad + \int_0^t \int_0^\tau \big((s^3 - \tau^3) X_{1,m-1}(\tau) + \tau^2(s^2 - \tau^2) X_{2,m-1}(\tau)\big) ds\, d\tau, \\ X_{2,m}(t) = \int_0^t \big(2(s-1) X_{1,m-1}(s) + (2s^4 + 2s^3 + 2s^2 - 1) X_{2,m-1}(s)\big) ds \\ \quad + \int_0^t \int_0^\tau \big((s^2 - \tau^2) X_{1,m-1}(\tau) + \tau^2(s^2 + \tau^2) X_{2,m-1}(\tau)\big) ds\, d\tau, \end{cases} \quad m = 2, 3, \ldots.$$

Now if we set $X_{1,1}(t) = 0$, then

$$(1-\alpha_0)t + \left(\frac{7\alpha_0}{2} - 3 - \frac{\alpha_1}{2}\right)t^2 + \left(\frac{7\alpha_1}{6} - 2\alpha_0 - \frac{\alpha_2}{3} + 1\right)t^3$$
$$+ \left(\frac{7\alpha_2}{12} + \frac{3\alpha_0}{4} - \frac{3\alpha_1}{4} - \frac{\alpha_3}{4} + \frac{1}{4} + \frac{\beta_0}{2}\right)t^4 + \left(\frac{7\alpha_3}{20} - \frac{2\alpha_2}{5} + \frac{3\alpha_1}{10} - \frac{\alpha_4}{5} + \frac{2\beta_0}{5} + \frac{\beta_1}{5} + \frac{1}{20}\right)t^5 + \cdots = 0,$$

and if we set $X_{2,1}(t) = 0$, then

$$-(1+\beta_0)t + \left(1 - \frac{\beta_1}{2} - \frac{\beta_0}{2} - \alpha_0\right)t^2 + \left(\frac{2\alpha_0}{3} - \frac{\alpha_1}{3} - \frac{\beta_1}{6} - \frac{\beta_2}{3} - \frac{1}{3}\right)t^3$$
$$+ \left(\frac{\alpha_1}{4} - \frac{\alpha_2}{6} + \frac{\beta_0}{2} - \frac{\beta_3}{4} - \frac{\beta_2}{12} + \frac{1}{3}\right)t^4 + \left(\frac{\beta_1}{5} + \frac{2\beta_0}{5} + \frac{2\alpha_2}{15} - \frac{\alpha_0}{20} - \frac{\beta_4}{5} - \frac{\alpha_3}{10} - \frac{\beta_3}{20} + \frac{1}{5}\right)t^5 + \cdots = 0.$$

It can be easily shown that

$$\alpha_0 = 1, \alpha_1 = 1, \alpha_1 = \frac{1}{2!}, \alpha_3 = \frac{1}{3!}, \alpha_4 = \frac{1}{4!}, \alpha_5 = \frac{1}{5!}, \ldots,$$
$$\beta_0 = -1, \beta_1 = 1, \beta_2 = -\frac{1}{2!}, \beta_3 = \frac{1}{3!}, \beta_4 = -\frac{1}{4!}, \beta_5 = \frac{1}{5!}, \ldots.$$

Therefore, the exact solutions of the system of integral-differential equations (11) can be expressed as

$$x_1(t) = 1 + \sum_{n=0}^{\infty} \frac{\alpha_n}{n+1} t^{n+1} = \sum_{n=0}^{\infty} \frac{t^n}{n!} = e^t,$$

$$x_2(t) = 1 + \sum_{n=0}^{\infty} \frac{\beta_n}{n+1} t^{n+1} = \sum_{n=0}^{\infty} (-1)^n \frac{t^n}{n!} = e^{-t}.$$

Example 2. Consider the following system of integro-differential equations with the exact solutions $x_1(t) = \cosh t$ and $x_2(t) = \sinh t$,

$$\frac{dx_1(t)}{dt} = -t^3 - 6t - 1 + x_1(t) + (7-2t)x_2(t)$$
$$+ \int_0^t \left((s+t)x_1(s) + (s-t)^3 x_2(s)\right) ds, \quad x_1(0) = 1,$$

$$\frac{dx_2(t)}{dt} = -3t^2 + t - 6 + (7-2t)x_1(t) + x_2(t) \qquad (15)$$
$$+ \int_0^t \left((s-t)^3 x_1(s) + (s+t)x_2(s)\right) ds, \quad x_2(0) = 0.$$

For solving system (15) by NHPM, we construct the following homotopy

$$\frac{dX_1(t)}{dt} = x_{1,0}(t) - p(x_{1,0}(t) + t^3 + 6t + 1 - X_1(t) - (7 - 2t)X_2(t)$$
$$- \int_0^t \Big((s+t)X_1(s) + (s-t)^3 X_2(s)\Big)ds\Big),$$
$$\frac{dX_2(t)}{dt} = x_{2,0}(t) - p(x_{2,0}(t) + 3t^2 - t + 6 - (7 - 2t)X_1(t) - X_2(t) \quad (16)$$
$$- \int_0^t \Big((s-t)^3 X_1(s) + (s+t)X_2(s)\Big)ds\Big).$$

Assuming that,

$$x_{1,0}(t) = \sum_{n=0}^{\infty} \alpha_n P_n(t), \; x_{2,0}(t) = \sum_{n=0}^{\infty} \beta_n P_n(t), \; P_i(t) = t^i, \; X_1(0) = 1, \; X_2(0) = 0.$$

Applying the inverse operator, $L^{-1} = \int_{t_0}^t (\cdot)\, dt$ to both sides of equation (16), we obtain

$$X_1(t) = 1 + \sum_{n=0}^{\infty} \frac{\alpha_n}{n+1} t^{n+1} - p\Big(\sum_{n=0}^{\infty} \frac{\alpha_n}{n+1} t^{n+1} + \frac{t^4}{4} + 2t^3 + t - \int_0^t \big(X_1(s) + (7 - 2s)X_2(s)\big) ds$$
$$- \int_0^t \int_0^\tau \big((s+\tau)X_1(s) + (s-\tau)^3 X_2(s)\big) ds\, d\tau\Big), \quad (17)$$
$$X_2(t) = \sum_{n=0}^{\infty} \frac{\beta_n}{n+1} t^{n+1} - p\Big(\sum_{n=0}^{\infty} \frac{\beta_n}{n+1} t^{n+1} + t^3 - \frac{t^2}{2} + 6t - \int_0^t \big((7 - 2s)X_1(s) + X_2(s)\big) ds$$
$$- \int_0^t \int_0^\tau \big((s-\tau)^3 X_1(s) + (s+\tau)X_2(s)\big) ds\, d\tau\Big).$$

Suppose the solutions of system (17) have the form (14), substituting (14) into (17) and equating the coefficients of p with the same power leads to

$$p^0 : \begin{cases} X_{1,0}(t) = 1 + \sum_{n=0}^{\infty} \frac{\alpha_n}{n+1} t^{n+1}, \\ X_{2,0}(t) = \sum_{n=0}^{\infty} \frac{\beta_n}{n+1} t^{n+1}, \end{cases}$$

$$p^1 : \begin{cases} X_{1,1}(t) = -\sum_{n=0}^{\infty} \frac{\alpha_n}{n+1} t^{n+1} - \frac{t^4}{4} - 3t^2 - t + \int_0^t \big(X_{1,1}(s) + (7-2s)X_{2,1}(s)\big) ds \\ \quad + \int_0^t \int_0^\tau \big((s+\tau)X_{1,1}(s) + (s-\tau)^3 X_{2,1}(s)\big) ds\, d\tau, \\ X_{2,1}(t) = -\sum_{n=0}^{\infty} \frac{\beta_n}{n+1} t^{n+1} - t^3 + \frac{t^2}{2} - 6t + \int_0^t \big((7-2s)X_{1,1}(s) + X_{2,1}(s)\big) ds \\ \quad + \int_0^t \int_0^\tau \big((s-\tau)^3 X_{1,1}(s) + (s+\tau)X_{2,1}(s)\big) ds\, d\tau, \end{cases}$$

$$p^m : \begin{cases} X_{1,m}(t) = \int_0^t \Big(X_{1,m-1}(s) + (7-2s)X_{2,m-1}(s)\Big)ds \\ \qquad + \int_0^t \int_0^\tau \Big((s+\tau)X_{1,m-1}(s) + (s-\tau)^3 X_{2,m-1}(s)\Big)ds\,d\tau, \\ X_{2,m}(t) = \int_0^t \Big((7-2s)X_{1,m-1}(s) + X_{2,m-1}(s)\Big)ds \\ \qquad + \int_0^t \int_0^\tau \Big((s-\tau)^3 X_{1,m-1}(s) + (s+\tau)X_{2,m-1}(s)\Big)ds\,d\tau, \end{cases} \quad m = 2, 3, \ldots.$$

If we set $X_{1,1}(t) = 0$, then

$$-\alpha_0 t + \left(\frac{\alpha_0}{2} + \frac{7\beta_0}{2} - 3 - \alpha_1\right)t^2 + \left(\frac{1}{2} + \frac{\alpha_1}{6} + \frac{7\beta_1}{6} - \frac{2\beta_0}{3} - \frac{\alpha_2}{3}\right)t^3$$
$$+ \left(\frac{\alpha_2}{12} + \frac{5\alpha_0}{24} + \frac{7\beta_2}{12} - \frac{\beta_1}{4} - \frac{1}{4} - \frac{\alpha_3}{4}\right)t^4 + \left(\frac{\alpha_3}{20} - \frac{\alpha_4}{5} + \frac{7\beta_3}{20} - \frac{2\beta_2}{15} + \frac{7\alpha_1}{120}\right)t^5 + \cdots = 0.$$

Further assume that $X_{2,1}(t) = 0$.

Then we have

$$(1-\beta_0)t + \left(\frac{7\alpha_0}{2} - \frac{1}{2} - \frac{\beta_0}{2} - \frac{\beta_1}{2}\right)t^2 + \left(\frac{7\alpha_1}{6} - \frac{2\alpha_0}{3} + \frac{\beta_1}{6} - \frac{\beta_2}{3} - 1\right)t^3$$
$$+ \left(\frac{7\alpha_2}{12} + \frac{5\beta_0}{24} - \frac{\beta_3}{4} - \frac{\alpha_1}{4} + \frac{\beta_2}{12}\right)t^4 + \left(\frac{7\alpha_3}{20} + \frac{7\beta_1}{120} + \frac{\beta_3}{20} - \frac{2\alpha_2}{15} - \frac{\beta_4}{5} - \frac{1}{20}\right)t^5 + \cdots = 0.$$

It can be easily shown that

$$\alpha_0 = 0, \alpha_1 = 1, \alpha_2 = 0, \alpha_3 = \frac{1}{3!}, \alpha_4 = 0, \alpha_5 = \frac{1}{5!}, \ldots,$$
$$\beta_0 = 1, \beta_1 = 0, \beta_2 = \frac{1}{2!}, \beta_3 = 0, \beta_4 = \frac{1}{4!}, \beta_5 = 0, \ldots.$$

Thus

$$x_1(t) = 1 + \sum_{n=0}^\infty \frac{\alpha_n}{n+1} t^{n+1} = \sum_{n=0}^\infty \frac{t^{2n}}{(2n)!} = \cosh t,$$
$$x_2(t) = \sum_{n=0}^\infty \frac{\beta_n}{n+1} t^{n+1} = \sum_{n=0}^\infty \frac{t^{2n-1}}{(2n-1)!} = \sinh t.$$

which are exact solutions.

CONCLUSION

In this work, we considered a new homotopy perturbation method for solving systems of linear integro-differential equations. New method is a powerful straightforward method. Using this method we obtained new efficient recurrent relations to solve these systems. The new homotopy perturbation method is apt to be utilized as an alternative approach to current techniques being employed to a wide variety of mathematical problems. The computations associated with the examples in this paper were performed using maple 10.

REFERENCES

[1] T.L. Bo, L. Xie, X.J. Zheng, Numerical approach to wind ripple in desert, *International Journal of Nonlinear Sciences and Numerical Simulation* 8 (2) (2007) 223–228.

[2] F.Z. Sun, M. Gao, S.H. Lei, et al., The fractal dimension of the fractal model of drop-wise condensation and its experimental study, *International Journal of Nonlinear Sciences and Numerical Simulation* 8 (2) (2007) 211–222.

[3] H. Wang, H.M. Fu, H.F. Zhang, et al., A practical thermodynamic method to calculate the best glass-forming composition for bulk metallic glasses, *International Journal of Nonlinear Sciences and Numerical Simulation* 8 (2) (2007) 171–178.

[4] L. Xu, J.H. He, Y. Liu, Electrospun nano-porous spheres with Chinese drug, *International Journal of Nonlinear Sciences and Numerical Simulation* 8 (2) (2007) 199–202.

[5] J.H. He, Homotopy perturbation technique, *Computer Methods in Applied Mechanics and Engineering* 178 (1999) 257–262.

[6] J.H. He, A coupling method of homotopy technique and perturbation technique for nonlinear problems, *International Journal of Non-Linear Mechanics* 35 (1) (2000) 37–43.

[7] J.H. He, Comparison of homotopy perturbation method and homotopy analysis method, *Applied Mathematics and Computation* 156 (2004) 527-539.

[8] J.H. He, Homotopy perturbation method: a new nonlinear analytical technique, *Applied Mathematics and Computation* 135 (2003) 73-79.

[9] J.H. He, The homotopy perturbation method for nonlinear oscillators with discontinuities, *Applied Mathematics and Computation* 151 (2004) 287-292.

[10] J.H. He, Application of homotopy perturbation method to nonlinear wave equations *Chaos, Solitons and Fractals* 26 (2005) 695–700.

[11] J.H. He, Homotopy perturbation method for solving boundary value problems, *Physics Letters* A 350 (2006) 87-88.

[12] J.H. He, Limit cycle and bifurcation of nonlinear problems, *Chaos, Solitons and Fractals* 26 (3) (2005) 827-833.

[13] J. Biazar, H. Ghazvini, Exact solutions for nonlinear Schrödinger equations by He's homotopy perturbation method. *Physics Letters* A 366 (2007) 79-84.

[14] D.D. Ganji, The application of He's homotopy perturbation method to nonlinear equations arising in heat transfer, *Physics Letters* A 355 (2006) 337–341.

[15] S. Abbasbandy, Numerical solutions of the integral equations: Homotopy perturbation method and Adomian's decomposition method, *Applied Mathematics and Computation* 173 (2006) 493-500.

[16] Z. Odibat, S. Momani, Modified homotopy perturbation method: application to quadratic Riccati differential equation of fractional order, *Chaos, Solitons and Fractals* [In press].

[17] A.M. Siddiqui, R. Mahmood, Q. K. Ghori, Homotopy perturbation method for thin film flow of a third grade fluid down an inclined plane, *Chaos, Solitons and Fractals* [In press].

[18] D.D. Ganji, A. Sadighi, Application of homotopy perturbation and variational iteration methods to nonlinear heat transfer and porous media equations, *Journal of Computational and Applied Mathematics* 207(2007)24-34.

[19] Golbabai, M. Javidi, Application of homotopy perturbation method for solving eighth-order boundary value problems, *Applied Mathematics and Computation*, 191 (2) (2007) 334-346.

[20] J.H. He, An elementary introduction to recently developed asymptotic methods and nanomechanics in textile engineering, *International Journal Of Modern Physics* B, 22(21)2008 3487-3578.

[21] Fatemeh Shakeri, Mehdi Dehghan, Solution of delay differential equations via a homotopy perturbation method, *Mathematical and Computer Modelling*, [In Press].

[22] Beléndez, T. Beléndez, A. Márquez, C. Neipp, Application of He's homotopy perturbation method to conservative truly nonlinear oscillators, *Chaos, Solitons and Fractals*, 37 (3) 2008 770-780.

[23] J. Biazar, H. Ghazvini, He's homotopy perturbation method for solving system of Volterra integral equations of the second kind, *Chaos, Solitons and Fractals* [In press].

[24] J. Biazar, H. Ghazvini, Numerical solution for special non-linear Fredholm integral equation by HPM, *Applied Mathematics and Computation*, 195 (2008) 681-687.

Chapter 10

NATURAL ADDITIVE FOR EOR SCHEME DURING CHEMICAL FLOODING AND ITS ENVIRONMENT FRIENDLY SUSTAINABLE APPLICATION

M. Safiur Rahman[*] and M. Rafiqul Islam

Environmental Engineering Program, Faculty of Engineering, Dalhousie University
D215-1360 Barrington Street, Halifax, Nova Scotia, Canada

ABSTRACT

Alkali and alkali/polymer solutions are well known techniques for the chemical flooding application. For this scheme, synthetic high-pH alkaline solutions are commonly used. These solutions are not environment friendly and are expensive. As a result, alkaline flooding has lost its appeal in last few decades. However, low-cost, environment-friendly alkaline solutions hold promises. This paper demonstrates how wood ash can be used as a source of low-cost alkali that is also environment friendly. The feasibility of using high pH alkaline solution, extracted from wood ash was conducted in the laboratory. From the experimental studies, it was found that the resulting solution was transparent and had high alkalinity. It was also found that the pH value of 6% wood ash-extracted solution was very close to the pH value of 0.5% synthetic sodium hydroxide or of 0.75% synthetic sodium meta silicate solution. A preliminary microscopic study of oil/oil droplets interaction in natural alkaline solution was carried out in order to understand the oil/water interface changes with time and its effect on oil/oil droplet coalescence. The microscopic study showed that two oil droplets were coalesced after 3.5 minute in 6% wood ash extracted solution. The interaction of the alkali in floodwater and the acids in reservoir crude oil result in the in-situ formation of surfactants that causes the lowering of interfacial tension (IFT) in caustic flooding that assist in the oil recovery process by mobilizing oil. In this study, the interfacial tension was measured using the Du Nouy ring method and it was observed that this environment friendly alkaline solution effectively reduced the interfacial tension with the acidic crude oil. Characterization of maple wood ashes has been was investigated using a variety of techniques, including, SEM-EDX, XRD, NMR. The SEM micrographs of the maple wood ash samples showed

---

[*] Corresponding author: safiur.rahman@dal.ca

that the ash samples consisted of some porous and amorphous particles of carbon and several inorganic particles of irregular shape. The X-ray analysis on maple wood ash revealed that the predominant elements in the wood ash samples were oxygen, calcium, potassium, silicon. Lesser amounts of the elements were sodium, magnesium, titanium and aluminum also observed in maple wood ashes. The XRD analysis revealed that the major components of maple wood ashes were calcium oxide, potassium oxide, manganese oxide, silica oxide and magnesium oxide which were alkaline in nature. The ^{13}C CP/MAS NMR spectrum of maple wood ashes showed a very pronounced intense peak around 168.36 ppm was revealed that was assigned for the carbonate, $[(O)_2-C=O]$. It was revealed that nutrient elements status in fresh and treated maple wood ashes is almost same. Therefore, after alkaline extraction for EOR application, the same maple wood ashes has potential to use as a source of nutrient to soil and plants.

Keywords: interfacial tension, wood ash, pH, organic acid, crude oil, NMR, SEM, EDX, XRD, alkalinity.

INTRODUCTION

Natural additives have been used for the longest time, dating back to the regime of the Pharaohs of Egypt and the Hans of China. However, the renaissance in Europe has given rise to industrial revolution that became the pivotal point for the emergence of numerous artificial chemicals. Today, thousands of artificial chemicals are being used in everyday product, ranging from health care products to transportation vehicles. With renewed awareness of the environmental consequences and more in-depth knowledge of science, we are discovering that such ubiquitous use of artificial chemicals is not sustainable (Khan, 2006). If the pathways of various artificial chemicals are investigated, it becomes clear that such chemicals cannot be assimilated in nature, making an irreparable footprint that can be the source of many other ecological imbalances (Islam, 2004; Chhetri et al., 2006; Chhetri and Islam, 2008). Federal regulators have determined that about 4,000 chemicals used for decades in Canada pose enough of a threat to human health or the environment that they need to be subjected to safety assessments (The Globe and Mail, 2006). These artificial additives are either synthetic themselves or derived through an extraction process that uses synthetic products.

Crude oil makes a major contribution to the world economy today. Crude oil development and production in oil reservoirs can include up to three distinct phases: primary, secondary, and tertiary (EOR) recovery. During the primary and secondary recovery, only 30% to 50% of a reservoir's original oil in place is typically produced (USDoE, 2006). Hence, attention is being paid to Enhanced Oil Recovery (EOR) techniques for recovering more oil from the existing oilfields. The worldwide target for EOR is estimated to be two trillion barrels. Alkaline flooding is one of the EOR recovery processes and it has begun in 1925 with the injection of a sodium carbonate "Soda ash" solution in the Bradford area of Pennsylvania (Mayer et al., 1983). The alkaline flooding process is simple when compared to other chemical floods, yet it is sufficiently complex to require detailed lab evaluation and careful selection of a reservoir for application. Caustic flooding is an economical option because the cost of caustic chemicals is low compared to other tertiary enhancement systems.

The chemicals most commonly used for alkaline flooding are sodium hydroxide (NaOH), sodium orthosilicate (Na_4SiO_4), sodium metasilicate (Na_2SiO_3), sodium carbonate (Na_2CO_3), ammonium hydroxide (NH_4OH), ammonium carbonate $(NH_4)_2CO_3$ (Jennings, 1975; Larrondo et al., 1985; Rahman, 2007). Due to reservoir heterogeneity and the mineral compositions of rock and reservoir fluids, the same alkaline solution might induce a different mechanism. A good number of laboratory investigations dealing with the interaction of alkaline solutions with reservoir fluids and reservoir rocks have been reported in the literature (Jennings, 1975; Ramakrishnan and Wasan, 1983; Trujillo, 1983). Due to higher pH value, sodium hydroxide is considered to be the most useful alkaline chemical for oil recovery schemes (Campbell, 1977). The price comparison of the most common synthetic alkaline substances between 1982 to 2006 is mentioned in Table 1. It shows that alkaline price has increased five to twelve times higher during the last fifteen years. The biggest challenge of any novel recovery technique is to be able to produce under attractive economic and environmental conditions (Islam, 1996; Khan and Islam, 2007; Chhetri et al., 2010). Due to the high cost of synthetic alkaline substances and the environmental impact, the alkaline flooding has lost its popularity. This is reflected in Figures 1 and 2.

Table 1. Comparison of price and physical properties of most common alkalis (Mayer et al., 1983; Chemistry Store 2005; ClearTech. 2006)

Name of alkali	Formula	pH of 1% solution	Na_2O (%)	Solubility ($gm/100cm^3$) Cold water	Solubility ($gm/100cm^3$) Hot water	Price range (dollar/ton) in 1988 (Mayer, et al. 1983)	Price range (dollar/ton) in 2006 (ClearTech 2006 Chemistry Store 2005)
Sodium Hydroxide	NaOH	13.15	0.775	42	347	285 to 335	830
Sodium Orthosilicate	Na_4SiO_4	12.92	0.674	15	56	300 to 385	1385
Sodium Metasilicate	Na_2SiO_3	12.60	0.508	19	91	310 to 415	1340
Ammonia	NH_3	11.45	-	89	7.4	190 to 205	1920
Sodium Carbonate	Na_2CO_3	11.37	0.585	7.1	45.5	90 to 95	1400

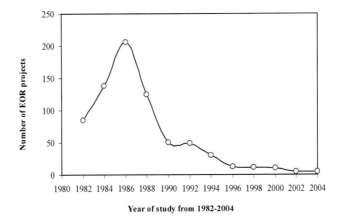

Figure 1. Total oil production by chemical flooding projects in the USA (Moritis, 2004).

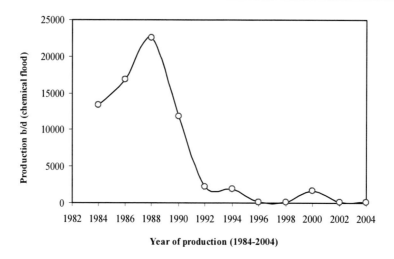

Figure 2. Chemical flooding field projects in the USA (Moritis, 2004).

These graphs have been generated using data reported by Moritis (Moritis, 2004). However, cost effective alkali might recover its popularity in the recovery scheme. It has become a research challenge for the petroleum industry to explore the use of low cost natural alkaline solutions for EOR during chemical flooding. In this paper, wood ash extracted solution is used as a low cost natural alkaline solution. Several experiments have been conducted to test the feasibility of that natural alkaline solution.

Mechanisms of Alkaline Flooding for EOR Scheme

Alkaline water flooding is an old recovery process in which pH of the injected water is increased by the addition of relatively inexpensive chemicals. Many crude oils naturally contain a certain amount of organic acids (Thibodeau et al., 2003; Islam and Farouq Ali, 1989). When this acidic oil is displaced by an alkaline solution, chemical reactions will occur at the interface. These reactions produce surface-active agents, which will in turn reduce the IFT between the two fluids, emulsify oil and water (Cooke et al., 1974; Islam, 1990). The reduction in interfacial tension between acidic oil and alkaline water depends on the pH of the water, the concentration and type of organic acids in the oil and the concentration and type of salts in solution Ramakrishnan and Wasan, 1983; Elkamel et al., 2002).

Some researchers (deZabala et al., 1982; Islam and Farouq Ali, 1989; Rahman et al., 2008) reported that there are four common mechanisms contribute to improve oil recovery with chemical methods. These mechanisms are (1) emulsification and entrainment; (2) wettability reversal water-wet to oil-wet, (3) wettability reversal oil-wet to water-wet; and (4) emulsification and entrapment. Other related mechanisms include emulsification with coalescence, wettability gradients, oil-phase swelling, disruption of rigid films and improved sweep resulting from precipitates altering flow (Campbell, 1977). All the postulated mechanisms have some superficial similarities. Laboratory experiments (Robinson et al., 1977) and field trials (Cooke et al., 1974) have shown that alkaline flooding performance will depend on: (1) water composition, (2) rock oil composition, (3) rock type and reactivity and

(4) alkaline concentration especially how it interacts with the previously mentioned parameters.

Toxicity of the Synthetic Alkalis

Alkali is one of the most commonly used chemicals for various applications. It has a wide range of application in different industries, such as petroleum refinery, pulp and paper mills, battery manufacturer, cosmetics, soap and detergent, leather processing industry, metal processing industry, water treatment plants etc. The estimated worldwide demand of sodium hydroxide was 44 million tons expressed as NaOH 100% in 1999. The global demand is expected to grow 3.1% per year (SAL, 2006).

In Figure 3, CMAI (2005) reported that 62 million tons of alkalis have been produced in 2005. Alkalis are raw commercial products, and when they are transferred to other parts of the manufacturer's plant for use in further chemical processing, there is always the risk of leakage. Each year huge amounts of synthetic chemicals are produced and all of the chemicals including all alkalis are considered responsible for direct or indirect pollution of the environment (Islam, 2006). These alkalis have also significant adverse effects on human health. Inhalation of dust, mist, or aerosol of sodium hydroxide and other alkalis may cause irritation of the mucous membranes of the nose, throat, and respiratory tract.

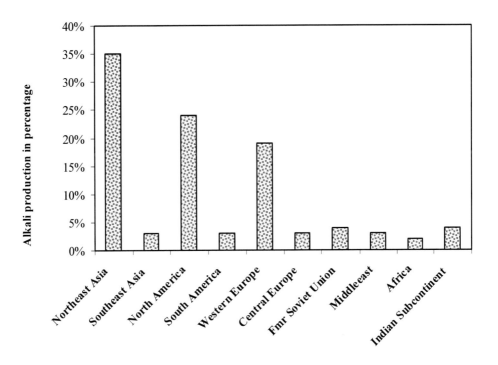

Figure 3. The total alkali production in the world in 2005 (CMAI, 2005).

Exposure to the alkalis in solid or solution can cause skin and eye irritation. Direct contact with the solid or with concentrated solutions causes thermal and chemical burns leading to deep-tissue injuries and also permanent damage to any tissue (MSDS, 2006;

ATSDR, 2006). Haynes (1976) reported that a dose of 1.95 grams of sodium hydroxide can cause death. Due to the high cost of synthetic alkaline substances and the environmental impact, alkaline flooding has lost its popularity. However, low-cost, environment-friendly alkaline solutions can hold promise (Rahman, 2007).

Alkalinity in Wood Ashes

Wood ash is a by-product of combustion in wood-fired power plants, paper mills, and other wood burning facilities. A huge amount of wood ash is produced every year worldwide and approximately three million tons of wood ash is produced annually in the United States alone (SAL, 2006). Wood ash is a complex heterogeneous mixture of all the non-flammable, non-volatile minerals which remain after the wood and charcoal have burned away. Because of the presence of carbon dioxide in the fire gases, many of these minerals will have been converted to carbonates (Dunn, 2003). The major components of wood ash are potassium carbonate 'potash' and sodium carbonate 'soda ash'. From a chemical standpoint these two compounds are very similar. From the 1700's through the early 1900's, wood was combusted in the United States to produce ash for chemical extraction. Wood ash was mainly used to produce potash for fertilizer and alkali for the industry. On an average, the burning of wood results in about 6-10% ashes. Ash is an alkaline material with a pH ranging from 9 – 13 (Rahman and Islam, 2008) and due to its high alkalinity characteristics, wood ash has various application in different sectors as an environment friendly alkaline substance.

MATERIALS AND METHODOLOGY

For our laboratory test, maple wood ash samples were collected from wood furnace and the ash samples were sifted with a sieve size of 30 to remove as much of charcoal as possible. The maple wood ash samples were characterized using variety of techniques i.e. scanning electron microscopy (SEM) couples with energy dispersive x-ray (EDX), x-ray diffraction analysis (XRD) and nuclear magnetic resonance (NMR). Screened maple wood ash samples and synthetic sodium hydroxide were taken in different amounts and placed in beakers. Different concentrated alkaline solutions (percentage of synthetic NaOH were 2.0%, 1.5%, 1.0%, 0.5%, 0.2% and percentage of maple wood ash samples were 8%, 6%, 4%, 2% and 1%) prepared for laboratory testing and alkalinity of the solutions was measured by a pH meter (Thermo Orion SP21, USA).

The alkalinity of the maple wood ash leachate was examined by making repeated batch extractions (Guenther, 1982) with deionized distilled water (L/S=10). A 5.0 g of maple wood ash was weighed into a 250ml of Erlenmeyer flask with an aliquot of 50 ml of deionized distilled water. The flask was then agitated on a wrist-action shaker for 1.0 h. After 1.0 h of agitation, the flask was kept constant for 10 min to settle the ash at bottom of the flask. The aqueous phase was separated by a pipette and pH was measured. Again a new aliquot (50 ml) of deionized distilled water was added into the Erlenmeyer flask. The procedure was repeated 10 single extraction.

Natural alkaline solutions at different concentrations were placed on microscopic slides and crude oil droplets were added with the help of a needle tip. The process of coalescences and flocculation of the oil droplets were observed by the Carl Zeiss light microscope attached with Axiovision 4.0 software and AxioCam digital camera in the petroleum laboratory of the Civil and Resource Engineering Department, Dalhousie University. Microscopic digital images of the crude oil were captured after every five seconds and the images of the oil droplets were analyzed using image-processing software. The summery of this study was shown in Figure 4 as a flow chat on the application of natural additive for enhanced oil recovery scheme.

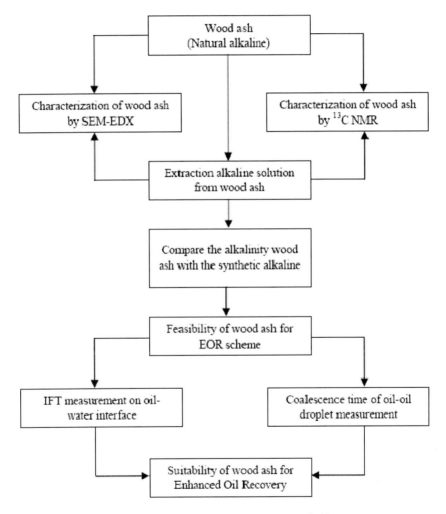

Figure 4. Major steps used to study the natural additives for enhanced oil recovery.

SEM-EDX Studies

Scanning electron microscopy is the best known and most widely-used of the surface analytical techniques. SEM, accompanied by X-ray analysis, is considered a relatively rapid, inexpensive, and basically non-destructive approach to surface analysis (Thipse et al, 2002).

Due to the manner in which the image is created, SEM images have a characteristic three-dimensional appearance and are useful for judging the surface structure of the samples. It is often used to survey surface analytical problems before proceeding to techniques that are more surface-sensitive and more specialized. SEM with or without energy or wavelength dispersive X-ray detectors for chemical analysis has been and still is routinely used in the study of rocks, minerals, biological samples. Several researchers used the SEM-EDX technique before to study the characterization of fly ash (Ayala et al., 1998; Erol et al., 2007). However, it was not studied for the analysis of maple wood ash for its characterization. The main objective of the SEM observation is to gather information on sample topography, crystallography and or elemental composition of maple wood ash samples. For the SEM and the microprobe facilities, the samples had to be coated in order to prevent charging the sample (Deydier et al., 2005).

Sample Preparation for SEM-EDX Studies

In order to view non-conductive samples, the SEM setup requires the samples to be electrically conductive (Hu et al., 2004). To perform our analysis, the maple wood ash samples were coated with a thin layer of conductive materials, namely, gold and palladium. Coating was performed with a small device called a sputter coater (Hitachi E-1030, Japan). The sputter coater uses argon gas and a small electric field. The samples were first fixed on double sided adhesive carbon tapes attached to one end of SEM stubs and then the samples were attached with carbon tapes.

Then the maple wood ash sample was placed in a small chamber which is at vacuum. Argon (Ar) was then introduced and an electric field was used to add a positive charge to the atoms. The Argon ions are then attracted to a negatively charged piece of gold and palladium foil. The Argon ions act like sand in a sandblaster, knocking gold and palladium atoms from the surface of the foil. Gold and palladium atoms now settle onto the surface of the sample, producing a gold-palladium coating. The thickness of the coating is 30.00nm, and the density is $19.32 g/cm^3$. The coated samples of maple wood ashes are investigated using a SEM (Hitachi S-4700, Japan) equipped with an Energy-dispersive X-ray Spectrometry (EDX) system (INCA, UK) in Institute of Research Materials (IRM), Dalhousie University. The experimental condition of SEM-EDX is given in Table 2.

Table 2. Experimental conditions of SEM-EDX for the analysis of maple wood ash

Particulars	Maple wood ash
Type	Default
Live time	100 sec
Real time	104.68 - 105 sec
Acquisition geometry (°)	Tilt = 0.0, Azimuth = 0.0
Spectrum processing	No peak omitted
Number of iteration	2-3
Processing option	All elements analyzed
Detector	Silicon

XRD Studies

The maple wood ash samples were analyzed by X-ray diffractometer (Siemens D5000) using a copper target to generate the X-rays (wavelength = 1.5405×10−10 m). The wood ash samples were first finely ground and then mounted on a sample holder. The mineralogy of the maple wood ash samples were determined by XRD using a continuous scan mode. The relative abundances of the individual mineral phases were based on the integrated intensity for each mineral phase.

The individual integrated peak intensities were summed to provide a total integrated intensity which was divided into the individual integrated intensities to provide a percentage of total integrated intensity. The relative abundances of the individual mineral phases were reported in terms of the percentage of total intensity represented by the individual mineral phases.

Nuclear Magnetic Resonance (NMR) Studies

NMR (Nuclear Magnetic Resonance) is a form of spectroscopy which works based on the absorption and emission of energy produced from changes in the spin states of the nucleus of an atom. In this study, ^{13}C CP/MAS NMR is used for powdered maple wood ash samples. The NMR test was conducted in the Chemistry Department, Dalhousie University, Canada. The ^{13}C NMR experiments were conducted on a Bruker Avance DSX NMR spectrometer with a 9.4 T magnet (proton Larmor frequency 400 MHz) using a HX probe head with rotors of 4mm diameter. The powdered maple wood ash samples were studied with ^{13}C cross polarization magic angle spinning (CP/MAS) NMR using TPPM proton decoupling and linearly varying proton contact powers.

The conditions for the pulse sequence are optimized on glycine, which also serves as external chemical shift reference with its carbonyl peak at 176.06 ppm. The cross polarization time is 2.6ms. The proton frequency for the CP and the proton decoupling is measured as the center frequency of the proton spectrum using single pulse excitation (Basu et al., 2007). Proton T_1 inversion/recovery measurements are used to determine the fastest possible repetition time for the ^{13}C cross-polarization experiments. The derived repetition times for the ^{13}C CP/MAS spectra was 0.3s for the maple wood ash samples. NMR Spectra were taken at 8.0 and 9.0 kHz sample spinning speeds to identify spinning sidebands but none are found. Each of the features in the spectra a functional group that resonates at the found ppm value, based on typical ^{13}C chemical shift (Hesse et al., 1979).

Computer Image Analyzer Studies

A microscopic study of oil/oil droplets interaction in wood ash extracted solution was carried out to understand the oil/water interface changes with time and its effects on oil/oil droplets coalescence. The images of oil/oil droplets interaction were taken by a professional camera scanner "Axio-Cam" with an optical microscope (Zeiss 459310). Axio-Vision has the ability to control, contrast, brightness and color of images as well as correction of lighting conditions and control of white balance, various processes for enhancing focus and emphasizing details-noise suppression shooting contour enhancement. The main objective of this software is to obtain as clear an image as possible for the microscope so it can be later analyzed using the KS300 software. KS300 is an image processing and analyzing software that permits fast and accurate measurement of morphological and densitometric parameters (Mustafiz, 2002). In addition, it has the ability to recognize with a very high sensitivity the contrast of different colors including gray, which allows us to study both black and white and colored images.

IFT Measurement Using Du Nouy Ring Method

Interfacial tension (IFT) is an important physical property. It characterizes interfaces between two immiscible liquids, frequently encountered and it has wide application in many industries (Alguacil, et al., 2006). Its value is usually determined by different methods, one of which is the Du Nouy ring method (Lecomte du Nou"y, 1919). In the ring method, the liquid is raised until contact with the surface is registered. The sample is then lowered again so that the liquid film produced beneath the ring is stretched. As the film is stretched, a maximum force is experienced which is recorded in the measurement.

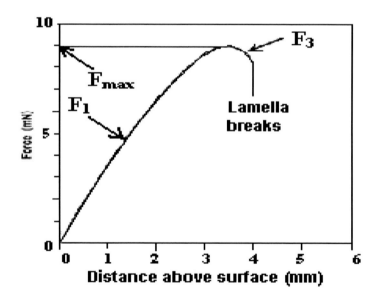

Figure 5. Change of force with ring distance (Redrawn from KRÜSS, 2006).

At the maximum of the force vector is exactly parallel to the direction of motion; at this moment the contact angle θ is 0 (see Figure 5). The following illustration shows the change of force as the distance of the ring from the surface increases.

In practice the distance is first increased until the area of maximum force has been passed through. The sample vessel containing the liquid is then moved back so that the maximum point is passed through a second time. The maximum force is only determined exactly on this return movement and used to calculate the tension. The calculation is made according to the following equation (KRÜSS, 2006).

$$\sigma = \frac{F_{max} - F_V}{L \cos \theta}$$

Here, σ = surface or interfacial tension; F max= maximum force; F V= weight of volume of liquid lifted; L= wetted length, θ = contact angle. The contact angle (θ), decreases as the extension increases and has the value 0° at the point of maximum force, this means that the term cos θ has the value 1.

RESULT AND DISCUSSION

Characterization of Maple Wood Ash Producing the Alkalinity

In this study, SEM-EDX was used to characterize the morphology and surface texture of individual particles of the maple wood ash samples and to determine the elemental composition presented in the samples. Each maple wood ash sample was characterized by randomly selecting 3–4 fields of view and examining all the ash particles observed within the selected fields. The elemental composition and morphology were noted for each particle and compiled for each sample. The morphology of the untreated maple wood ash samples using the magnification of 200 μm revealed that the maple wood ash samples consisted of some irregular shaped amorphous particles and porous particles (Figure 6).

After alkaline materials extraction from the maple wood ash samples SEM-EDX analysis was carried to observe the changes in its structures. The SEM micrograph using the magnification of 50μm revealed that the treated maple wood ash sample (Figure 7) was more dense than the untreated sample due to chemical reaction of the wood ash components in aqueous solution. To get the elemental composition in the particles of the untreated and the treated maple wood ash samples, Energy-Dispersive X-ray (EDX) microanalysis was carried out in this study.

Figure 6. SEM image of untreated maple wood ash sample.

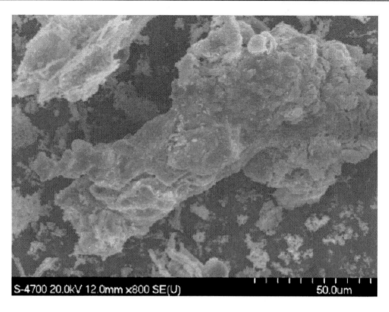

Figure 7. SEM image of maple wood ash sample after the treatment.

Table 3. Elemental analysis of maple wood ash samples by EDX coupled with SEM

Serial No	Elements in wood ashes	Before the extraction of alkaline solution		After the extraction of alkaline solution	
		Atomic (%)	Weight (%)	Atomic (%)	Weight (%)
01.	Oxygen (O)	86.60	73.85	86.22	73.54
02.	Sodium (Na)	0.54	0.66	0.46	0.56
03.	Magnesium (Mg)	1.31	1.69	1.29	1.67
04.	Aluminum (Al)	0.75	1.07	1.09	1.56
05.	Silicon (Si)	0.90	1.35	1.42	2.12
06.	Potassium (K)	1.55	3.24	1.44	2.99
07.	Calcium (Ca)	7.92	16.92	7.70	16.45
08.	Titanium (Ti)	0.14	0.35	0.12	0.29
09.	Manganese (Mn)	0.30	0.87	0.28	0.81
10.	Total		100.00		100.00

The EDX detector is capable of detecting elements with an atomic number equal to or greater than six. The intensity of the peaks in the EDX is not a quantitative measure of elemental concentration, although relative amounts can be inferred from relative peak heights. EDX coupled to SEM analysis of ash showed that the predominant elements in the wood ash samples were oxygen, calcium, potassium, silicon, and aluminum. Lesser amounts of the elements were sodium, magnesium and titanium observed in the untreated maple wood ash and treated maple wood ash samples (Table 3; Figured 8 and 9).

It was revealed from the SEM-EDX analysis that the elemental composition and nutrients for plants in untreated maple wood ashes and treated maple wood ashes were almost same. However, after alkaline solution extraction from the wood ashes, those wood ashes might be used as a source nutrients for plants growth.

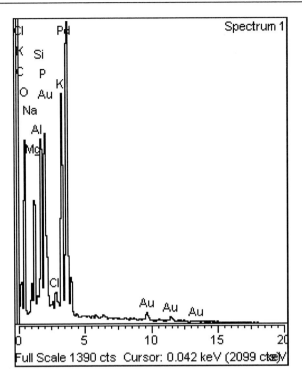

Figure 8. Corresponding EDX coupled with SEM spectrums of untreated maple wood ash sample (Spectrum-1).

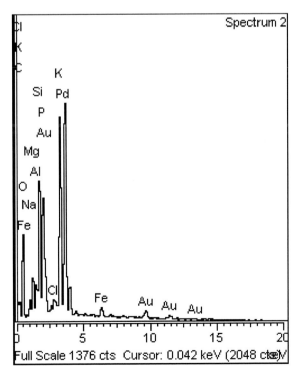

Figure 9. Corresponding EDX coupled with SEM spectrums of treated maple wood ash sample (Spectrum-2).

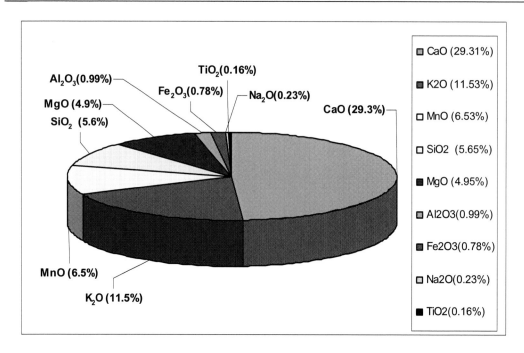

Figure 10. Mineralogical composition of maple wood ash sample.

The major compounds in maple wood ashes, identified by XRD diffraction, are displayed in Figure 10. The XRD analysis revealed that the major components of maple wood ash were calcium oxide, potassium oxide, manganese oxide, silica oxide and magnesium oxide which were alkaline in nature (Holmberg et al., 2003; Rahman and Islam, 2008). The XRD observations were also consistent with energy dispersive x-ray (EDX) analysis coupled with SEM data (Table 3) on selected samples of maple wood ashes. During the combustion of wood, organic compounds are mineralized and the basic cations are transformed to their oxides which are slowly hydrated and subsequently carbonated under atmospheric conditions. The crystalline compounds were found to contain mainly Ca, Mg and in a lesser extend K in all ashes studied. The mineralogical speciation of maple wood ash showed that calcium occurs in a variety of compounds. Very soluble forms such as calcium oxide (CaO) and portlandite (Ca(OH)$_2$) dominate, but calcite (CaCO$_3$) with low solubility occur in significant amounts (Steenari et al., 1999).

The ^{13}C CP/MAS NMR spectrum of maple wood ash sample spun at 8.0 kHz, together with the experimental parameters and peak frequencies are shown in Figure 11. A very pronounced feature in the maple wood ash spectra was an intense peak around 168.363 ppm was revealed that was assigned for the carbonate, [(O)$_2$-C=O] based on typical ^{13}C chemical shift tables (Hesse *et al.,* 1979; Adelaide, 2006). These carbonate ions presented in maple wood ash reacted with whatever species were around. If there is a lot of hydrogen ion (acidic solution), they stick to the carbonate ion and form new ions, namely, *hydrogen carbonate* ions, or *bicarbonate* ions. If there is not much hydrogen ion around, carbonate in effect 'steals' a hydrogen ion from water, leaving a hydroxide ion behind and producing an alkaline solution (Dunn, 2003). Normally water does not ionize but, in presence of carbonate and biocarbonate ions, it also breaks apart into ions [H$_2$O$_{(l)}$ → H$^+_{(aq)}$ + OH$^-_{(aq)}$], such as ionic compounds, which consequently increase the alkalinity of the aqueous solution. It can be observed in the Figure 10 that the major components in wood ash are CaO (29.3 %) and K_2O (11.5 %). These two compounds produce alkalinity in an aqueous solution (Holmberg et al., 2003) and the following reactions were happened during the process. The same types of mechanisms have been proposed earlier by several researchers (Steenari and Lindqvist, 1997; Dunn, 2003; Rahman and Islam, 2008) for the alkaline solution extraction from carbonate salts.

$$CaO + H_2O \rightarrow Ca(OH)_2;\qquad K_2O + H_2O \rightarrow KOH$$

$$Ca(OH)_2 + CO_2 \rightarrow CaCO_{3(s)};\qquad KOH + CO_2 \rightarrow K_2CO_{3(s)}$$

$$CaCO_{3(s)} \leftrightarrow Ca^{2+} + CO_3^{2-}{}_{(aq)};\qquad K_2CO_{3(s)} \leftrightarrow Ca^{2+}{}_{(aq)} + CO_3^{2-}{}_{(aq)}$$

$$H_2O_{(l)} \rightarrow H^+{}_{(aq)} + OH^-{}_{(aq)} \text{ (in presence of carbonate ions)}$$

$$CO_3^{2-}{}_{(aq)} + H^+{}_{(aq)} \leftrightarrow HCO_3^-{}_{(aq)}$$

$$HCO_3^-{}_{(aq)} + H_2O_{(l)} \leftrightarrow H_2CO_3^-{}_{(aq)} + OH^-{}_{(aq)}$$

$$CO_3^{2-}{}_{(aq)} + H_2O_{(l)} \rightarrow HCO_3^-{}_{(aq)} + OH^-{}_{(aq)}$$

Renton and Brown (1995) reported that the production of alkalinity depends on the relative amounts of the individual alkaline components. The dissolution of CaO and $Ca(OH)_2$ results in a high initial production of alkalinity while the dissolution of $CaCO_3$ results in an overall increase in the amount of alkalinity generated and an increase in the longevity of alkaline production. It is observed in Figure 10 that the main component of maple wood ash is CaO 29.3%.

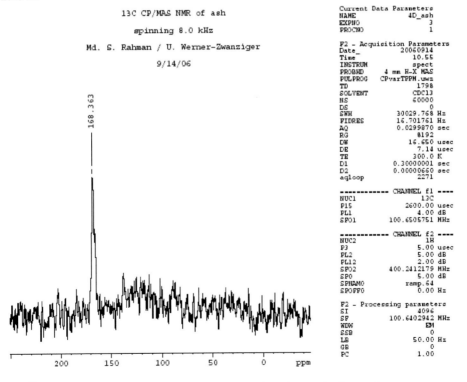

Figure 11. ^{13}C CP/MAS NMR high resolution spectrum of maple wood ash sample.

This value is very close to the reported value by the several authors (Renton and Brown, 1995; Obernberger et al., 1997) for the different types of ashes such as fluidized bed combustor ash (12.2-29.5 % CaO), straw ash (7.8 % CaO), cereals ash (7.1 % CaO), wood bark ash (42.2 % CaO) and wood chips (44.7 % CaO) are used as an alkali. However, wood ash might have the potential to produce solutions of high alkalinity depending on the CaO and $CaCO_3$ contents.

Alkalinity of Maple Wood Ash Extracted Solution

Different amounts of maple wood ash samples were taken in different beakers and each beaker contained the same amount (100 ml) of double distilled water. The shaker was used to mix the maple wood ash samples properly in water. The prepared maple wood ash extracted solutions were filtered with Whatman-40 filter paper. Before filtering the wood ash solution was dark, however after filtration the maple wood ash extracted solution was clear (Figure 12). The pH values of maple wood ash extracted solutions at different percentages of maple wood ash (1 %, 2 %, 4 %, 6 % and 8 %) were measured and the pH values were presented in Table 4. It is found that the alkalinity (pH value) of 6 % wood ash solution is close to 0.5 % synthetic sodium hydroxide solution. This value is also very close to the pH value of 0.75 % Na_2SiO_3 solution (Rahman et al., 2008). It was reported in literature that the alkaline solution of pH range 12 – 14 was treated as a "strong base", as shown in Figure 13. From the experimental studies it was revealed that 4-8 % wood ash extracted solution might have potentiality to use as a source strong natural alkaline solution and it might be used on enhanced oil recovery (EOR) scheme during chemical flooding.

Figure 12. After filtration of natural alkaline solution (6% maple wood ash sample).

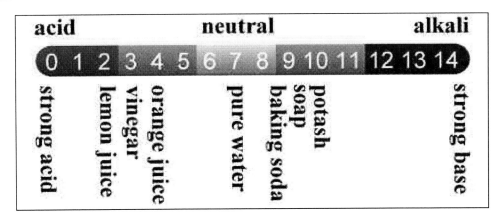

Figure 13. A Typical pH scale (Caveman Chemistry, 2006).

During the alkaline flooding, pH value of the synthetic alkaline maintains in the range of 11.5 – 13.5 as a common practice. Therefore, it might be proposed that the natural alkaline solution extracted from the 6% wood ash could be used instead of 0.5% synthetic sodium hydroxide solution or 0.75% synthetic sodium metasilicate solution during the chemical flooding scheme in an acidic reservoir. Burk (1987) has reported that Na_2CO_3 solutions are less corrosive to sandstone than NaOH or Na_4SiO_4. The buffering action of sodium carbonate (Na_2CO_3) can reduce alkali retention in the rock formation. The main composition of wood ash is carbonate slats such as $CaCO_3$, Na_2CO_3 (soda ash) and K_2CO_3 (potash). Carbonate slats offer an additional advantage upon contact with hard water. The resulting carbonate precipitation does not adversely affect permeability as compared to the precipitations of the hydroxides or silicate (Rahman et al., 2008). It is, therefore, suggested that the use of carbonate buffer solution extracted from maple wood ash might result in longer alkali breakthrough times and increased tertiary oil recovery during chemical flooding.

Table 4. Comparison of alkalinity between natural alkaline solution extracted from wood ash and synthetic sodium hydroxide solution at different concentrations

Synthetic sodium hydroxide solution		
	Synthetic NaOH solution (%)	pH value
Synthetic sodium hydroxide solution (NaOH)	2.0% NaOH solution	13.11
	1.5% NaOH solution	13.05
	1.0% NaOH solution	12.74
	0.5% NaOH solution	*12.35*
	0.2% NaOH solution	11.95
Wood ash Solutions		
	Percentage of wood ash solution	pH value
Maple wood ash solution	8% wood ash solution	12.42
	6% wood ash solution	*12.29*
	4% wood ash solution	12.09
	2% wood ash solution	11.83
	1% wood ash solution	11.42

The alkalinity of the ash leachate after repeated batch extraction with deionized distilled water at L/S = 10 is illustrated in Figure 14. It was reveled that the pH value is above 10.85 after 10 times repeated extractions due to presence of different alkaline components present is wood ashes. Therefore, same wood ashes might be used for alkaline solution extraction for several times.

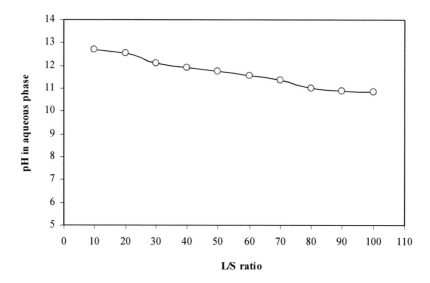

Figure 14. pH of the ash leachate as a function of L/S ratio.

Feasibility Test of Maple Wood Ash Extracted Solution for EOR Applications

A series of laboratory experiments was conducted on the natural alkaline solution for its application in chemical flooding using crude oil. The physical properties of the crude oil are given in Table 5 (Rahman et al., 2008). A microscopic study of oil/oil droplets interaction in maple wood ash extracted solution was carried out to understand the oil/water interface changes with time and its effects on oil/oil droplets coalescence. When the oil droplet was added to the natural alkaline solution, alkali reacted with the organic acids of oil. As a result, surfactant was produced.

Table 5. Physical properties of the crude oil

Sl No	Physical properties	Value
01.	Specific gravity	: 0.7 to 0.95
02.	Vapor pressure	: > 0.36 Kpa at 20°C
03.	Vapor density	: 3 to 5 (approx),
04.	Freezing point	: – 60°C to –20°C
05.	Viscosity,	: < 15 centistokes at 20°C
06.	Solubility	: Insoluble
07.	Co-efficient of water/oil distribution	: <1

This surfactant contained hydrophilic molecules and hydrophobic molecules that started to form a layer around the oil droplet called "micelles" and it caused the smoothing of surfaces, resulting in reduced interfacial friction. Once the micelle is formed, mobility of oil droplets increases and oil droplets move faster under the influence of buoyancy force or viscous force which result in drainage of thin surfactant water film at the contact between flocculating oil droplets (Figure 15). Consequently, this film reaches the critical thickness at which it ruptures and oil droplets coalesce to form a larger globule shown in Figures 16A through 16F. It was also found that two oil droplets coalesce after 3.5 minutes in 6 % wood ash solution that contain the same alkalinity of 0.5 % NaOH and 0.75 % Na_2SiO_3 solutions.

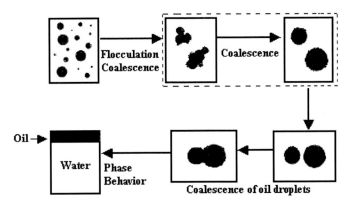

Figure 15. Schematic illustration of different steps in oil droplets growth during coalescence.

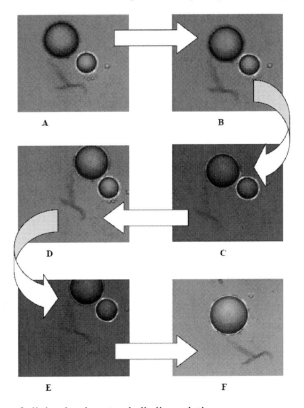

Figure 16. Coalescence of oil droplets in natural alkaline solution.

Interfacial Tension (IFT) Measurement

IFT measurements between a crude oil and an alkaline solution have generally been accepted as a screening tool to evaluate the EOR potential of the crude oil by alkali (Jennings, 1975; Campbell, 1977, deZabala, et al., 1982). Recently, Mollet et al. (1996) showed in an experimental study that minimum IFT is not observed in absence of alkali in the aqueous phase. From our experimental studies, it is found that IFT gradually decreases with increasing concentration of natural alkaline solutions (Figure 17) as well as with increasing concentration of NaOH solutions (Figure 18).

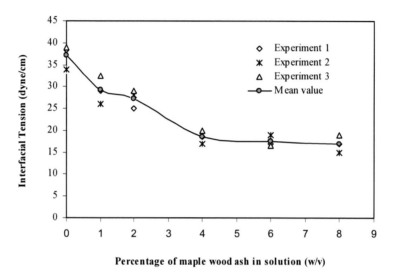

Figure 17. Interfacial tension vs different concentration of wood ash solution at 22°C (experimental results are plotted showing the points and the mean values for the triplicate analysis are used to draw the trend line).

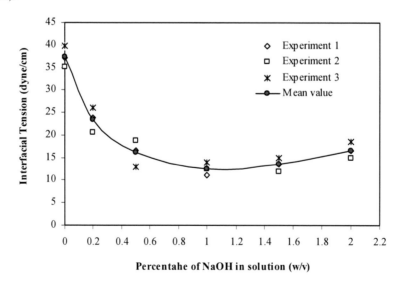

Figure 18. Interfacial tension vs pH of NaOH solutions at 22°C (experimental results are plotted showing the points and the mean values for the triplicate analysis are used to draw the trend line).

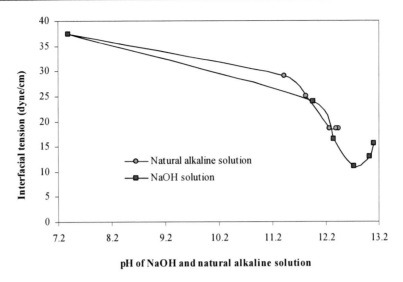

Figure 19. IFT of crude oil vs pH of NaOH and natural alkaline solutions at 22°C (mean values for the triplicate analysis are used to draw the graph).

It was observed that IFT decreases up to a certain limit with pH values, which was illustrated in Figure 19. This behavior is typical of dynamic interfacial phenomena, which are known to take place for heterogeneous fluids (Elkamel et al., 2002). The higher concentration of the alkaline solution develops more surface active agents as a result of the reaction between organic acid in the crude oil and alkali in the aqueous phase. Hence, this surface active agent (petroleum soap) can cause the decrease of interfacial tension and increase the mobility of oil in the continuous water phase.

Sustainable Process on Wood Ashes in Environment

In Canada 4,175,000 km² land out of total 10 million square kilometers land is covered by forest. Every year a huge amount of wood ash is produced worldwide and approximately three million tons of wood ash is produced annually in the United States alone (SAL, 2006). From the 1700's through the early 1900's, wood was combusted in the United States to produce ash for chemical extraction. Wood ash was mainly used to produce potash for fertilizer and alkali for the industry. It also has great potential to be used as a source of major and micro-nutrient elements required for healthy plant growth. Wood ash contains many essential nutrients mainly calcium, potassium, magnesium, and phosphorus for the growth of trees and other plants. Therefore, wood ashes might have potential applications in different sectors as an environment friendly sustainable natural additive.

Rahman et al. (2004) studied maple wood ash as an environment friendly adsorbent without any physical and chemical treatment to remove both arsenic (III) and arsenic (V) from contaminated aqueous streams at low concentration levels. Arsenic removal upto 80% was observed following statistic test and it was shown that the arsenic concentration was reduced from 500ppb to 5ppb following dynamic column tests. Chetri and Islam (2008) reported a process that could render the biodiesel production process tuely green. The conventional catalysist to produce biodisel is mainly used synthetic alkali (sodium hydroxide) that is very toxic for human as well for the environment. The catalysist used by the authors to produce biodiesel is natural alkali extracted from the wood ashes that could hold the promises to overcome the limitation of the existing toxic and synthetic chemicals. It has been reported in literature (Anfiteatro, 2007) that wood ashes with neem seed oil, kefir, sea salt and essential oils might be used to make a natural toothpaste, which may help with tooth sensitivity due to the poor condition of tooth enamel, or for the prevention of sensitive teeth. This toothpaste has also shown to eliminate bleeding of the gums, if the paste is used on a daily basis. The tooth paste can remove most stains on the teeth, such as those caused by cigarette smoking. It may also strengthen tooth enamel and gums. Apart from the traditional usage of wood ash as a source of alkali to the different sectors, it has been also used for a long time to saponify of fats in soap making, shampoo producing (Sh, 2007). Chhetri et al. (2009) has developed a process to produce completely natural birth soap using all natural ingredients such as vegetable oil, coconut oil, olive oil, honey, beeswax, cinnamon powder, neem leaf powder, natural coloring and flavoring agents instead of synthetic materials. Wood ashes extracted alkaline solution has been used for saponifying the oils to make the natural soap by the authors.

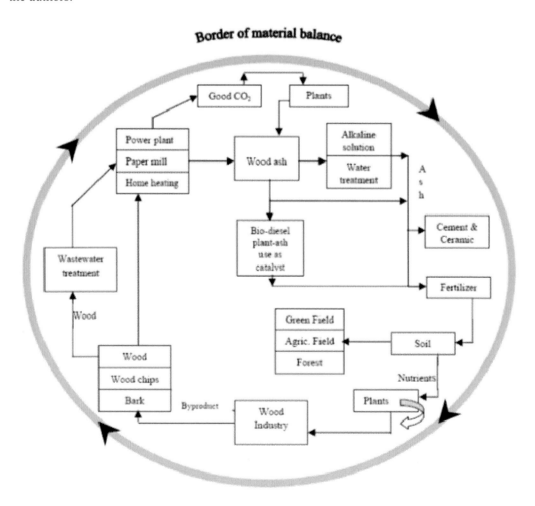

Figure 20. Possibilities for a sustainable utilization of wood ashes.

From this study, it was revealed that the nutritional quality in wood ashes after alkaline solution extraction and before extraction from wood ashes was almost same. However, the wood ash could be collected separately from different sources and it could be disposed to the landfill as the nutrients source of plants; or industrially utilized to cement manufacture, glazing agent in the ceramics industry, road base, puzzolona, alkaline material for the neutralization of wastes (Liodakis, 2005) and to contribute to the establishment of a sustainable process showing in Figure 20.

CONCLUSION

Based on the experimental results presented in this work, the following conclusions can be reached. This study supports the basic idea of the applicability of natural alkaline solution for the EOR scheme. However, a wide range of experiments from different viewpoint may increase this appealing field application.

- The natural alkaline solution extracted from the maple wood ash is highly alkaline and the alkalinity (pH value) of 6% wood ash solution is very close to 0.5% synthetic sodium hydroxide and 0.75% synthetic sodium metasilicate solution.
- Maple wood ash extracted alkaline solution reduces the interfacial tension with crude oil, which helps to increase oil mobility in aqueous phase.
- Coalescence time of oil droplets seems to be strongly influenced by the early micelles forming stage. It is also dependent on the decrease in interfacial tension. From the study, it is observed that coalescence time of oil/oil droplets decreases with the increasing pH and two oil droplets are coalesced after 3.5 minute in 6% maple wood ash solution. Maple wood ash extracted alkaline solution, which contains mainly soda ash (Na_2CO_3) and potash (K_2CO_3) could be more advantageous than the alkaline solution of NaOH or Na_4SiO_4 for alkaline flooding, because the buffered slug would be less reactive with sandstone minerals due to reduction of hydroxyl ion activity.
- Maple wood ash extracted alkaline substances are environment friendly and naturally abundant; where as injected synthetic alkaline solutions are cost effective and environmentally toxic and harmful.
- Characterization of maple wood ashes using SEM-EDX, XRD, NMR has been revealed that it also has great potential to be used as a source of major and micro-nutrient elements required for healthy plant growth and the nutrient elements status in treated maple wood ashes are almost same. However, after alkaline extraction, the same wood ashes might be used as liming and nutrient source to the solid and then plants.

ACKNOWLEDGMENT

The authors would like to thank the Atlantic Canada Opportunities Agency (ACOA) for funding this project under the Atlantic Innovation Fund (AIF).

REFERENCES

Adelaide, (2006). Waite Solid-state NMR Facility, University of Adelaide, Australia <http://www.waite.adelaide.edu.au/NMR/nmrlist.html> (accessed: December 30, 2006).

Alguacil, D.M., Fischer, P., and Windhab, E.J. (2006). Determination of the Interfacial Tension of Low Density Difference Liquid–Liquid Systems Containing Surfactants by Droplet Deformation Methods, *Chemical Engineering Science*, Vol. 61, pp. 1386-1394.

Anfiteatro, D.N. (2007). Dom's Tooth-Saving Paste, <http://users.chariot.net.au/~dna/toothpaste/toothpaste.htm> (accesssd: March 22, 2007).

ATSDR (Agency for Toxic Substances and Disease Registry), (2006). Medical Management Guide Line for Sodium Hydroxide, <http://www.atsdr.cdc.gov/MHMI/mmg178.html> (accessed: June 07, 2006).

Ayala, J., Blanco, F., Garcia, P., Rodríguez, P. and Sancho, J. (1998). Asturian Fly Ash as a Heavy Metals Removal Material, *Fuel*, Vol. 77(11), pp. 1147-1154.

Basu, A., White, R.L., Lumsden, M.D., Bisop, P., Butt, S., Mustafiz, S. and Islam, M.R. (2007). Surface Chemistry of Atlantic Cod Scale, *J. Nature Science and Sustainable Technology*, Vol. 1(1), pp. 69-78.

Burk, J.H. (1987). Comparison of Sodium Carbonate, Sodium Hydroxide, and Sodium Orthosilicate for EOR, *SPE Reservoir Engineering*, Vol. 2, pp. 9-16.

Campbell, T. C. (1977). A Comparison of Sodium Orthosilicate and Sodium Hydroxide for Alkaline Waterflooding, *Journal for Petroleum Technology*, SPE 6514, pp. 1-8.

Caveman Chemistry (2006). <http://cavemanchemistry.com/browse.html> (accessed: July22, 2006).

Chemistry Store (2005). <http://www.chemistrystore.com/index.html> (accessed: June 07, 2006).

Chhetri, A.B. and Islam, M.R. (2008). 'Towards Producing Truly Green Biodiesel', *Energy Sources*, Vol. 30, Issue 8, pp. 754-764.

Chhetri, A.B. and Islam, M.R. (2008). Inherently Sustainable Technology Development. Nova Science Publisher, New York.

Chhetri, A.B., Rahman, M.S. and Islam, M.R. (2006). Production of Truly 'Healthy' Health Products, 2nd Int. Conference on Appropriate Technology, July 12-14, Zimbabwe.

Chhetri, A.B., Rahman, M.S. and Islam, M.R. (2009). Characterization of Truly 'Healthy' Health Products, *Journal of Characterization and Development of Novel Materials*, Vol. 1, Issue 1, pp. 1-12.

Chhetri, A.B., Watts, K.C., Rahman, M.S. and Islam, M.R. (2010). Soapnut extract as natural surfactant for enhanced oil recovery, *Energy Sources, Part A (Recovery, Utilization and Environmental Effect)* (USA), Vol. 31(20), p. 1904-1914

ClearTech, (2006). Industrial Chemicals, North Corman Industrial Park, Saskatoon S7L 5Z3, Canada, <http://www.cleartech.ca/products.html> (accessed: May08, 2006).

CMAI, (2005). Chemical Market Associates Incorporated, <www.kasteelchemical.com/slide.cfm> (accessed: May 20, 2006).

Cooke, Jr., C. E., Williams, R. E., and Kolodzie, P. A. (1974). Oil Recovery by Alkaline Water Flooding, *J. Pet. Technol.*, Vol. 26, pp. 1356–1374.

deZabala, E. F. and Radke, C. J. (1982). The Role of Interfacial Resistances in Alkaline Water Flooding of Acid Oils, paper *SPE* 11213 presented at the 1982 SPE Annual Conference and Exhibition, New Orleans, 26-29.

Deydier, E., Guilet, R., Sarda, S., and Sharrock, P. (2005). Physical and Chemical Characterisation of Crude Meat and Bone Meal Combustion Residue: "Waste or Raw Material?" *Journal of Hazardous Materials*, Vol. 121(1-3), pp. 141–148.

Dunn, K. (2003). Caveman Chemistry, Chapter-8, Universal Publishers, USA.

Elkamel, A., Al-Sahhaf, T. and Ahmed, A.S. (2002). Studying the Interactions Between an Arabian Heavy Crude Oil and Alkaline Solutions, *Journal Petroleum Science and Technology*, Vol. 20(7), pp. 789–807.

Erol, M., Kucukbayrak, S. and Ersoy-Mericboyu, A. (2007). Characterization of Coal Fly Ash for Possible Utilization in Glass Production, *Fuel*, Vol. 86, pp. 706–714.

Guenther, W. B. (1982). Wood Ash Analysis: An Experiment for Introductory Courses. *J. Chem. Educ.*, Vol. 59, pp. 1047-1048.

Haynes, H.J., Thrasher, L.W., Katz, M.L. and Eck, T.R. (1976). Enhanced Oil Recovery, National Petroleum Council. An Analysis of the Potential for Enhanced Oil Recovery from Known Fields in the United States.

Hesse, M., Meier, H. and Zeeh, B. (1979). Spektroscopische Methoden in Der Organischen Chemie, Thieme VerlagStuttgart.

Holmberg, S.L., Claesson, T., Abul-Milh, M. and Steenari, B.M. (2003). Drying of Granulated Wood Ash by Flue Gas from Saw Dust and Natural Gas Combustion, *Resources, Conservation and Recycling*, Vol. 38, pp. 301-316.

Hu, P.Y., Hsieh, Y.H., Chen, J.C. and Chang, C.Y. (2004). Characteristics of Manganese-Coated Sand Using SEM and EDAX Analysis, *J. Colloid and Interface Sc.* Vol. 272, pp. 308-313.

Islam, M. R. (1990). New Scaling Criteria for Chemical Flooding Experiments, *Journal of Canadian Petroleum Technology*, Vol. 29(1), pp. 30-36.

Islam, M. R. (1996). Emerging Technologies in Enhanced Oil Recovery, *Energy Sources*, Vol. 21, pp. 97-111.

Islam, M.R. (2004). Unraveling the Mysteries of Chaos and Change: The Knowledge-Based Technology Development, *EEC Innovation*, Vol. 2(2), pp. 45-87.

Islam, M.R. (2006). A Knowledge-Based Water and Waste-Water Management Model. International Conference on Management of Water, Wastewater and Environment: Challenges for the Developing Countries, September 13-15, Nepal.

Islam, M.R. and Farouq Ali, M.S. (1989). Numerical Simulation of Alkaline/Cosurfactant/Polymer Flooding; Proceedings of the UNI-TAR/UNDP Fourth Int. Conf., Heavy Crude and Tar Sand.

Jennings, H.Y. Jr. (1975). A Study of Caustic Solution-Crude Oil Interfacial Tensions, *Society of Petroleum Engineers Journal*, SPE-5049, pp. 197-202.

Khan, M. I., Islam, M. R. (2007). True Sustainability in Technological Development and Natural Resource Management, Nova Science Publishers, USA.

Khan, M.I. (2006). Towards Sustainability in Offshore Oil and Gas Operations, Ph.D. Dissertation, Department of Civil and Resource Engineering, Dalhousie University, Canada, 440 pp.

KRÜSS (2006). Instruments for Surface Chemistry, Measuring Principle of KRÜSS Tensio Meters, KRÜSS GmbH, Wissenschaftliche Laborgeräte, Hamburg, Germany.

Larrondo, L.E., Urness, C.M. and Milosz, G.M. (1985). Laboratory Evaluation of Sodium Hydroxide, Sodium Orthosilicate, and Sodium Metasilicate as Alkaline Flooding, *Society of Petroleum Engineering,* Vol. 13577, pp. 307-315.

Lecomte Du Noüy, P. (1919). A new apparatus for measuring surface tension, *J. Gen. Physiol.,* Vol. 1, pp. 521–524.

Liodakis, S., Katsigiannis, G., Kakali, G. (2005). Ash Properties of Some Dominant Greek forest species, *Thermochimica Acta,* Vol. 437, pp. 158-167.

Mayer, E. H., Berg, Carmichael, R. L. and Weinbrandt, R. M. (1983). Alkaline Injection for Enhanced Oil Recovery - A Status Report, *Journal of Petroleum Technology,* Vol. 35, pp. 209-221.

Mollet, C., Touhami, Y., Hornof, V. (1996). A Comparative Study of the Effect of Ready Made and in-Situ Formed Surfactants on IFT Measured by Drop Volume Tensiometry, *J. Colloid Interface Sci.* Vol. 178: 523.

Moritis, G. (2004). Point of View: EOR Continues to Unlock Oil Resources, *Oil and Gas Journal.* ABI/INFORM Global, Vol. 102(14), pp. 45-49.

MSDS (Material Safety Data Sheet). (2006). Canadian Centre for Occupational Health and Safety, 135 Hunter Street East, Hamilton ON Canada L8N 1M5.

Mustafiz, S. (2002). A Novel Method for Heavy Metal Removal from Aqueous Streams, MASc Dissertation, Dalhousie University, Canada.

Obernberger, I., Biedermann, F., Widmann, W. and Riedl, R. (1997). Concentrations of Inorganic Elements in Biomass Fuels and Recovery in the Different Ash Fractions, *Biomass and Bioenergy,* Vol. 12(3), pp. 211-224.

Rahman, M.H., Wasiuddin, N., and Islam, M.R. (2004). Experimental and Numerical Modeling Studies of Arsenic Removal with Wood Ash from Aqueous Streams, *Can. J. Chem. Eng.*, Vol. 82(5), pp. 968–977.

Rahman, M.S. (2006). Effect of Natural Alkaline Solution on EOR During Chemical Flooding, Project Report, Department of Civil and Resource Engineering, Dalhousie University, Canada.

Rahman, M.S. (2007). The Prospect of Natural Additives in Enhanced Oil Recovery and Water Purification Operation, M.A.Sc. Dissertation, Dalhousie University, Canada.

Rahman, M.S. and Islam, M.R. (2008). Physico-Chemical Characterization of Ashes from *Acer nigrum* by SEM, XRD and NMR Technique, *J. Nature Science and Sustainable Technology*, Vol. 2 Issue 4, pp. 493-504.

Rahman, M.S., Hossain, M.E. and Islam, M.R. (2008). An Environment-Friendly Alkaline Solution for Enhanced Oil Recovery, *J. Petroleum Science and Technology,*Vol. 26, pp. 1596-1609.

Ramakrishnan, T.S. and Wasan, D.T. (1983). A Model for Interfacial Activity of Acidic Crude Oil-Caustic Systems for Alkaline Flooding, *SPE Journal*, SPE-10716, pp. 602-618.

Renton, J.J. and Brown, H.E. (1995). An Evaluation of Fluidized Bed Combustor Ash as a Source of Alkalinity to Treat Toxic Rock Materials, *Engineering Geology,* Vol. 40, pp. 157-167.

Robinson, R. J., Bursell, C.G. and Restine, J, L. (1977). A Caustic Steam flood Pilot-Kern River Field, paper SPE 6523 presented at SPE AIME 47th Annual California Regional Meeting, 13akerafield,California, April 13-15.

SAL, (2006). Soil Acidity and Liming: Internet Inservice Training, Best Management Practices for Wood Ash Used as an Agricultural Soil Amendment, <http://hubcap.clemson.edu/~blpprt/bestwoodash.html> (accessed: June 7, 2006).

Sh. A.M.M. (2006). Murugappa Chettiar Research Centre, Photosynthesis and Energy Division, Tharamani, Madras- 600 113, India.

Steenari, B.M. and Lindqvist, O. (1997). Stabilisation of Biofuel Ashes for Recycling to Forest soil., *Biomass and Bioenergy;* Vol. 13(1-2), pp. 39-50.

The Globe and Mail, Saturday 27 May 2006, Page A4.

Thibodeau, L., Sakanoko, M. and Neale, G. H. (2003). Alkaline Flooding Processes in Porous Media in the Presence of Connate Water, *Powder Technology,* Vol. 32, pp. 101–111.

Thipse, S. S., Schoenitz, M., and Dreizin, E. L. (2002). Morphology and Composition of the Fly Ash Particles Produced in Incineration of Municipal Solid Waste, *Fuel Processing Technology,* Vol. 75(3), pp. 173– 184.

Trujillo, E. M. (1983). The Static and Dynamic Interracial Tensions between Crude Oils and Caustic Solutions, *SPEJ* 23, pp. 645-656.

USDoE. (2006). <http://www.fossil.energy.gov/programs/oilgas/eor/index.html>, US Department of Energy (accessed : April 06, 2006).

In: New Developments in Sustainable Petroleum Engineering ISBN: 978-1-61324-159-2
Editor: Rafiq Islam © 2012 Nova Science Publishers, Inc.

Chapter 11

CHARACTERIZATION AND SEPARATION OF OIL-IN-WATER EMULSION

Ajay Mandal[*,1], *Pradeep Kumar,*[1] *Keka Ojha*[1] *and Subodh K. Maiti*[2]

[1]Department of Petroleum Engineering
Indian School of Mines, Dhanbad, India
[2]Department of Environment Science and Engineering
Indian School of Mines, Dhanbad, India

ABSTRACT

Chemical demulsification is most widely applied in petroleum industries, painting and wastewater treatment technology and involves the use of chemical additives to accelerate the emulsion breaking process. The stability of the emulsion has been characterized and it is observed that the stability depend on oil-water contact time, turbulence and amount of oil in contact with the water. A series of experiments have been carried out with different demulsifiers for separation of oil from oil-in-water emulsion. More than 90% separations are obtained with some demulsifiers under specific operating conditions.

Keywords: oil-in-water emulsion, demulsification, chemicals, COD.

1. INTRODUCTION

Oil exploitation is always accompanied by varying quantities of oily wastewater. The studies on stability and demulsification of crude oil emulsions have been reported by many investigators (Auflem et al., 2001; Li at al., 2007; Qiao et al., 2008).

---

[*] Corresponding Author: E-mail: mandal_ajay@hotmail.com

Other sources of oil-containing wastewaters include petroleum refineries, metal fabrication plants, rolling mills, chemical processing plants, machine shops, and vehicle maintenance shops (Webb, 1991; Bennett, 1998). Many different types of oils may be present in oily wastewater such as crude oil, diesel fuel, cutting and grinding oils, lubricating oils, water soluble coolants, natural animal or vegetable fats or any other organic immiscible in water.

The removal of these oily wastes from wastewater is of importance in preventing pollution and meeting environmental compliance standards. Oily waste removal may also be beneficial for water and oil recovery and reuse. Disposed water should contain less than 40 mg/l of oil (Moosai et al, 2003) and this requirement is becoming more enforced as damaging environmental effects from oily wastewater become more apparent.

Oil in wastewater mostly remains in the form of emulsion (Rupesh et al, 2008). Emulsions are suspensions of droplets, greater than 0.1 μm, consisting of two completely immiscible liquids, one of which is dispersed throughout the other. Their kinetic stability is a consequence of small droplet size and the presence of an interfacial film around water droplets. An emulsifying agent must be present to form stable oil-in-water emulsions (Sjoblom et al., 1992). Such agents include clay particles, added chemicals or the crude oil components like asphaltenes, waxes, resins and napthenic acids. These stabilizers suppress the mechanism involved in emulsion breakdown. The most important characteristics of oil-in-water emulsion are its stability (Fingas et al., 2004). The four classes of stability are: stable, meso-stable, unstable and entrained water. Out of these four states only stable and meso-stable can be characterized as emulsions.

The separation of oil from oil-water emulsion may be achieved by different methods, viz. ultracentrifugation (Hahn and Vold, 1975), coagulation (Zhang et al, 2005), flocculation (Bratskaya et al., 2006), synergistic effect of salt and ozone (Song et al., 1998)], electric field (Tsuneki et al., 2004), air flotation (Al-Shamrania et al., 2002) and membrane filtration (Kocherginsky et al., 2003, Chen et al., 2009). However, chemical demulsification is the most widely applied method of treating oil-in-water emulsions and involves the use of chemical additives to accelerate the emulsion breaking process. The advantages of chemical demulsification method are rapid water drop rate, better interface definition, better water quality, low temperature performance and overall improved cost efficiency for demulsification operations. The formulation of an emulsion demulsifier for specific oil emulsion is a complicated undertaking. The demulsification using chemicals is a very complex phenomenon. Demulsifier displace the natural stabilizers present in the interfacial film around the water droplets. The displacement is brought about by the adsorption of the demulsifiers at the interface. This disoplacement, occurring at the oil-water interface, influences the coalescence of water droplets through enhanced film drainage. The efficiency of demulsifier is dependent on its adsorption at the oil-water or droplet surface.

The best demulsifiers are those that rapidly displace preformed rigid films and leave a mobile film in its place. In petroleum system, asphaltenes and resinous substances comprise a major portion of the inetrfacially active components of the oil (Sjoblom et al., 1992; Johansen et al, 1989; Urdahl et al., 1992). Commercial demulsifiers are generally polymeric surfactants such as copolymers of polyxyethylene and polypropylene or alkylphenol-formaldehyde resins or blends of various surface-active substances (Taylor, 1992; Kokal, 2002).

Due to more and more severe environmental constraints, there is now a strong need in the oil production to restrict the use of chemicals and to utilize safer formulations, less toxic but at least as efficient as classical demulsifiers.

The present work conducted to study the influence of different chemical demulsifiers on the destabilization of emulsions and hence separation of oil from wastewater. The effects of time and temperature on the demulsification were also studied.

2. EXPERIMENTAL

2.1. Materials

In this study, six different demulsifiers; Polyethylene Glycol-400, Polyethylene Glycol-4000, Polyethylene Glycol 6000, Polyethylene oxide, N-hexylamine Amine, N-octylamine were used for demulsification of oil-in-water emulsion. Gear oil (EPX 90) available in the market with sp. Gravity 0.905 and kinematic viscosity 197 cSt at 40°C and 17.3 cSt at 100°C was used for preparation of emulsion in distilled water.

2.2. Methods

Oil-in-water emulsion was prepared at room temperature with a standard three blade propeller at speed of various rpm. For oil-in-water emulsion distilled water was used as the continuous phase and oil was used as the dispersed phase. The emulsion is then left to stand in separation flask for 6 hours and the bottom part is separated out. The separation flask can easily lead to the separation of two different density fluids based on their density. For preparation of emulsion, the oil water mixtures were stirred at different rpm and time. It has been found that the stability and hence the characteristics of emulsion is very much dependent on the stirring speed and time. Emulsions are then analyzed and treated with different demulsifiers for demulsification. The viscosity of the emulsion was measured in viscometer (Brookfield DV II+), microscopic picture of emulsion was taken by Olympus BX-60 Microscope and density by conventional method. The Chemical Oxygen Demand (COD) of the emulsion is considered as the oil equivalent in the emulsion. Thus COD of the sample were measured before and after treatment of the emulsion to determine the degree of separation of oil from emulsion. Typical characteristics of four different sets of emulsion have been shown in Table 1.

3. RESULTS AND DISCUSSION

3.1. Characterization of Emulsion

The concentration of oil in the oil-in-water emulsion is dependent on the presence of excess oil during formation of emulsion. A series of experiments were performed to characterize the emulsion formation as shown in Figure 1.

From the figure, it may be seen that the emulsification of oil (COD) in water first increases with increase in concentration of oil due to significant interaction between oil and water. However, after a certain oil to water ratio the coalescence of dispersed oil droplets increases in presence of excess oil and hence concentration of oil in the emulsion gradually decreases and then stabilizes at constant value.

The stirring speed has a strong effect on the emulsion formation. It has been found (Figure 2) that the concentration of emulsified oil in water first increases and then decreases due to breaking of emulsion at higher stirring speed.

Table 1. Typical characteristic of oil-water emulsions

Sample	1	2	3	4
Stirring speed (RPM)	2184	3963	5145	6870
Mixing time (min)	120	120	120	120
pH	7.8	7.96	7.78	7.40
Surface tension (Dynes/cm)	34	33	33	32
Viscosity (cp) at 25°C	2.83	4.19	6.45	4.45
Conductivity, mho	0.385	0.456	0.422	0.472
COD (mg/lit)	24320	27200	33920	30400

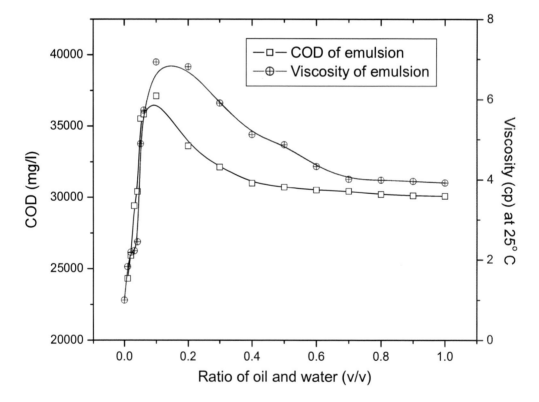

Figure 1. Effect of oil/water ratio on the stability of emulsion.

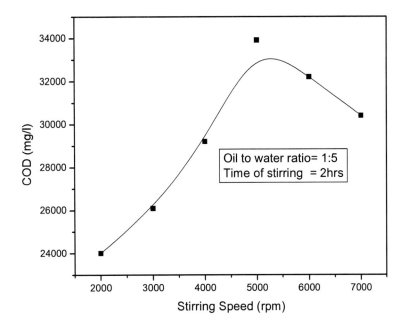

Figure 2. Effect of stirring speed on stability of emulsion.

3.2. Demulsification of Emulsion

In order to investigate the effect of different demulsifier and corresponding dosage on the demulsification of an oil-in-water emulsion, a series of experiments were carried out with different demulsifier. The experiments were performed at a temperature of 30°C and 1440 minutes retention time was provided for separation. Figure 3 shows the influence of polyethylene oxide on destabilization of emulsion.

The removal of oil increases with increase in amount of polyethylene oxide added due to significant reduction of interfacial tension between oil and water. However after a certain concentration the interfaces are saturated with the demulsifier and stagnation in oil and water separation are occurred. From the Figure 3 one may see that maximum separation occurs at around 1400 ppm of polyethylene oxide added.

Figure 4 shows the effect of different amine and glycol on the separation. The plots show the same trend as that of polyethylene glycol, but percentage separation of oil significantly higher particularly with N-Octylamine, N-Hexyl amine and Polyethylene Glycol-400. The separation capacity of N-Octyl amine (68.39%) was found to be higher than N-Hexyl amine (59.26 %) at room temperature due to higher molecular weight which acts as flocculants in adsorption and interaction activities (Abdurahman et al., 2007).

Polyethylene Glycol-400 was found to have better separation efficiency than Polyethylene Glycol-4000 and 6000 and similar trend was also reported by Abdurahman et al. (2007).

The retention time after addition of demulsifier in the emulsion was found to have a strong effect on separation. A minimum retention time should be provided flocculation and adsorption of oil particle. The Figure 5 shows that a minimum 120 minutes retention time should be provided for significant separation.

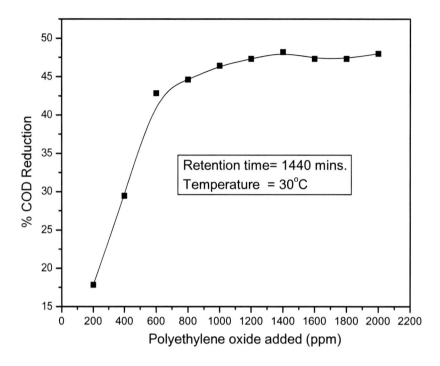

Figure 3. The influence of Polyethylene oxide on separation of oil from emulsion.

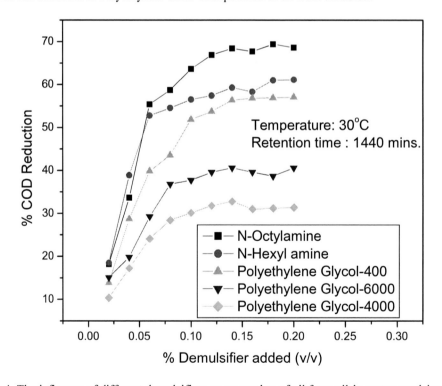

Figure 4. The influence of different demulsifiers on separation of oil from oil-in-water emulsion.

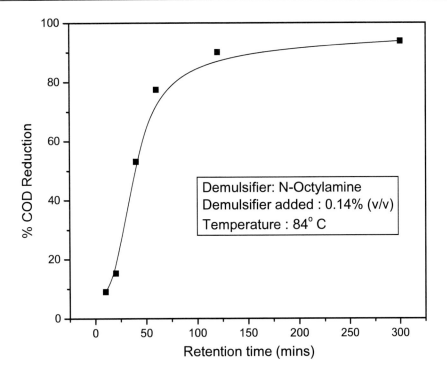

Figure 5. Effect of retention time on separation of oil from oil-in-water emulsion.

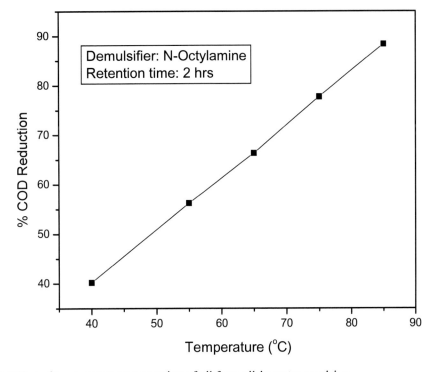

Figure 6. Effect of temperature on separation of oil from oil-in-water emulsion.

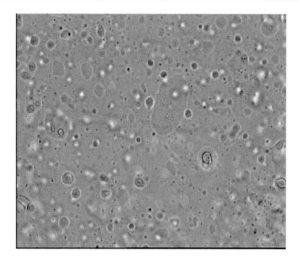

Figure 7. Microscopic picture (Magnification 430x) of untreated oil-in-water emulsion.

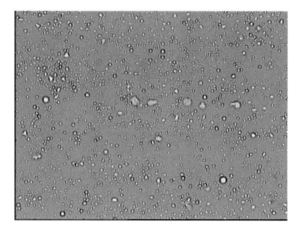

Figure 8. Microscopic picture (Magnification 430x) of oil-in-water emulsion treated with Polyethylene Glycol 400.

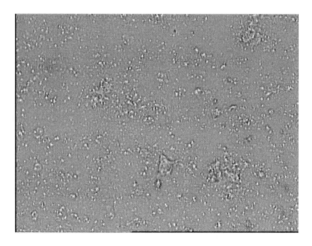

Figure 9. Microscopic picture (Magnification 430x) of oil-in-water emulsion treated with N-Octylamine.

Figure 6 shows the effect of treating temperature on the separation of emulsion. The removal of oil increases with the treating temperature. It is generally believed that the Brownian movement of molecules or finely grounded particles is responsible for this response. Heating up not only intensifies Brownian motion of the particles in oil emulsion to make them collided and coagulated effectively, also falls the internal viscosity of emulsion body vastly to weaken therefore the stability of emulsion.

Table 2. Analysis of microscopic pictures of emulsions

Untreated emulsion (Figure 7)	
Min. Val.	6.955 μm
Max. Val.	121.709 μm
Mean	20.760 μm
Std. Dev	18.478
Sum	830.409
No. of Samples	40
Emulsion treated with Polyethylene Glycol 400 (Figure 8)	
Min. Val.	1.832 μm
Max. Val.	21.871 μm
Mean	6.478 μm
Std. Dev	3.762
Sum	278.562
No. of Samples	43
Emulsion treated with N-Octylamine (Figure 9)	
Min. Val	0.918 μm
Max. Val.	8.237 μm
Mean	3.116 μm
Std. Dev	2.138
Sum	101.077
No. of Samples	40

Microscopic picture of oil-in-water emulsion is shown in Figure 7 while that of demulsifier treated samples are shown in Figure 8 and 9 respectively. The figures show the size distribution of droplets of different samples. From the figures it may be seen that the size of the emulsified water droplet in untreated emulsion is higher compared to the demulsifier treated emulsion. A detailed size distributions are shown in Table 2.

CONCLUSION

Based on the experimental study it can be concluded that the stability of emulsion is dependent on the different influencing conditions of the formation, viz. formation time, degree of turbulence and oil to water ratio. Demulsification of the emulsion were studied using different demulsifiers, out of which N-octylamine, N-Hexylamine and Polyethylene glycol-400 were found to have good removal efficiency of oil from water. The results suggest that addition of around 0.15 % (v/v) liquid demulsifiers give the optimum removal of oil from the emulsion. A minimum retention time of 120 minutes after addition of demulsifiers should be provided for better separation. The separation efficiency was also found to increase with increasing treating temperature.

REFERENCES

Abdurahman, H. N., Yunus, R. M. and Jemaat, Z. (2007) 'Chemical demulsification of water-in-crude oil emulsion', *J. Appl. Science*, Vol. 7, pp 196-201.

Al-Shamrania, A.A., James, A., Xiao, H. (2002) 'Separation of oil from water by dissolved air flotation, *Colloid Surf. A*, Vol. 209, pp 15–26.

Auflem, I. H., Kallevik, H, Westivik, A, Sjoblom, J. (2001) 'Influence of pressure and solvency on the separation of water-in-crude-oil emulsions from the North Sea', *Journal of Petroleum Science and Engineering*, Vol. 31, pp 1–12.

Bennett, G.F. (1988) 'The removal of oil from wastewater by air flotation', *CRC Critical Reviews in Environmental Control*, Vol.18, pp 189–253.

Bratskaya, S., Avramenko, V., Schwarz, S., Philippova, I. (2006) 'Enhanced flocculation of oil-in-water emulsions by hydrophobically modified chitosan derivatives,*Colloid Surf. A,* Vol.275, pp 168–176.

Chen, W., Peng, J., Su, Y., Zheng, L., Wang, L., Jiang, Z. (2009) 'Separation of oil/water emulsion using Pluronic F127 modified polyethersulfone ultrafiltration membranes', *Separation and Purification Technology*, Vol. 66, pp 591-597.

Fingas, M. and Fieldhouse, B. (2004) 'Formation of water-in-oil emulsions and application to oil spill modelling', *Journal of Hazardous Materials*, Vol. 107, pp. 37-50.

Hahn, A.U., Vold, R.D. (1975) 'The kinetics and mechanism of ultracentrifugal demulsification', *J. Colloids Interface Sci.* Vol. 51, pp 133–142.

Johansen, E. J., Skjarov, I. M., Lund, T. and Sjoblom, J. (1989) 'Water-in-Crude oil emulsions from Norwegian continental shelf. Part 1. Formation, Characterization and Stability correlations', *Colloids and Surfaces*, Vol.34, pp. 353-370.

Kocherginsky, N.M., Tan, C.L., Lu, W.F.(2003) 'Demulsification of water-in-oil emulsions via filtration through a hydrophilic polymer membrane, *J. Membr. Sci.* Vol. 220, 117–128.

Kokal, S. (2002) 'Crude oil emulsions: a state of the art review', *In Proceedings SPE ATCE*, San Antonio, Tx, SPE paper no 77497, pp 1-11.

Li, X.B., Liu J.T., Wang Y.T., Wang C.Y., Zhou, X.H. (2007) 'Separation of Oil from Wastewater by Column Flotation', *J China Univ Mining and Technol*, Vol. 17, pp 546 – 551.

Moosai, R. L., Dawe A. R. (2003) 'Gas attachment of oil droplets for gas flotation for oily wastewater cleanup', *Separation and Purification Technology*, Vol. 33, pp. 303-314.

Qiao, X., Zhang, Z. Yu, J. Ye, X. (2008) 'Performance characteristics of a hybrid membrane pilot-scale plant for oilfield-produced wastewater', *Desalination*, Vol. 225, pp 113–122.

Rupesh M. B., Prasad, B., Mishra I.M. and Kailas L. (2008) 'Wasewar Oil field effluent water treatment for safe disposal by electroflotation', *Chemical Engineering Journal,* Vol. 137, pp 503-509.

Sjoblom, J., Mingyuan, L., Christy A. A. and Gu, T. (1992) 'Water-in-Crude oil emulsions from Norwegian continental shelf 7', *Colloids and Surfaces*, Vol. 66, pp. 55-62.

Song, Y.C., Kim, I.S., Koh, S.C. (1998) 'Demulsi.cation of oily wastewater through a synergistic effect of ozone and salt, *Water Sci. Technol.* Vol. 38, pp 247–253.

Taylor, S.E., (1992) 'Resolving crude oil emulsions', *Chemistry and Industry*, Vol. 20. pp. 770-773.

Tsuneki, I., Keisuke, I., Shigeki, Y. (2004) 'Rapid demulsi.cation of dense oil-in-water emulsion by low external electric field, *Colloid Surf. A*, Vol. 242, pp 21–26.

Urdahl, O., Brekke, T. and Sjoblom, J. (1992) '13 C n.m.r and multivariate statistical analysis of adsorbed surfaceactive crude oil fractions and the corresponding crude oils', *Fuel*, Vol. 71, pp 739-746.

Webb, C. (1991) 'Separating oil from water', *The Chemical Engineer*, Vol. 11, pp 19–22.

Zhang, Z.Q., Xu, G.Y., Wang, F., Dong, S.L., Chen, Y.J. (2005) 'Demulsification by amphiphilic dendrimer copolymers', *J. Colloids Interface Sci.* Vol. 282, pp 1–4.

INDEX

A

accessibility, 83
acid, 142, 161
acidic, x, 141, 144, 154, 157
activation energy, 22
additives, x, 142, 147, 169, 170
adsorption, 170, 173
adverse effects, 145
aesthetic, 125
age, 55
algorithm, 57, 84, 85, 86, 87, 88, 89, 90, 91, 92, 95, 96, 97, 98
alkaline flooding, x, 141, 142, 143, 144, 146, 157, 163
alkalinity, x, 141, 142, 146, 154, 155, 156, 157, 158, 159, 163
amine, 173
ammonium, 143
annealing, 85, 87, 88, 90, 97, 98
APL, 37
aquifers, ix, 36, 55, 63, 90, 92, 98, 109, 110, 128
argon, 148
arsenic, 161
Artificial Neural Networks, 88
assessment, viii, 2, 5, 10, 11, 35, 36, 37, 39, 41, 44, 46, 47, 48, 49, 50, 51, 91, 110, 111, 124, 127, 128, 129
atmosphere, 57
atoms, 148
attachment, 178
Austria, 76, 107
authorities, 36
awareness, 142

B

barium, 100
barium sulphate, 100
base, 2, 11, 55, 71, 80, 85, 90, 103, 156, 163
behaviors, 36
Beijing, 109, 127
benzene, ix, 43, 45, 109, 111, 112, 116, 117, 118, 120, 121, 122, 124, 125, 126, 127
benzene, toluene, and ethylbenzene (BTE), ix, 109
bias, 90
bicarbonate, 154
biodegradation, 113, 114, 116, 117
biodiesel, 161
biological markers, 111
biological processes, 80
biological samples, 148
bioremediation, 61, 89, 90, 92, 94, 95, 97, 98
bleeding, 162
blends, 170
blood, 111
body weight, 115
boundary value problem, 132, 139, 140
bounds, 37, 39
breakdown, 170
breeding, 84
Brownian motion, 177

C

calcium, x, 142, 152, 154, 161
calculus, 85
calibration, 80, 98
cancer, ix, 37, 110, 112, 116, 121, 122, 123, 126, 127
capillary, 54, 56, 115
carbon, x, 114, 125, 126, 127, 142, 146, 148

carbon dioxide, 146
carcinogen, 115, 125
Caribbean, 76, 107
case studies, 85
case study, ix, 2, 49, 50, 55, 63, 92, 96, 110, 128
cation, 179
caustic flooding, x, 141
CBS, 5
challenges, 14
chemical, viii, x, 65, 100, 111, 112, 115, 128, 141, 142, 143, 144, 145, 146, 148, 149, 151, 154, 156, 157, 158, 161, 169, 170, 171
chemical demulsification, x, 169, 178
chemical properties, 111, 112
chemical reactions, 144
chemicals, ix, 109, 110, 115, 142, 143, 144, 145, 162, 169, 170, 171
China, 109, 127, 142, 178
chitosan, 178
chlorinated hydrocarbons, 51
cigarette smoking, 162
city, 77
classes, 90, 170
classification, 66, 100
cleanup, 50, 81, 86, 88, 92, 95, 178
coconut oil, 162
collaboration, 128
color, 149
combustion, 146, 154
commercial, 145
communities, 42, 47
compaction, 31
complexity, ix, 82, 87, 109
compliance, 10, 170
composition, 2, 116, 139, 144, 148, 151, 152, 154, 157
compounds, 110, 116, 128, 146, 154
compressibility, 17, 19, 21, 23, 27, 57
computation, 5, 26, 27, 83, 84, 85
computational grid, 43
computational performance, 90
computer, 55
computing, 14, 50
condensation, 139
conductivity, 58, 59, 64, 87, 89, 113, 115, 117
conference, 128
configuration, 3, 10, 114
conservation, 38, 57, 64
constant rate, 67
constituents, 37, 125
construction, 82
contact time, x, 169

contaminant, vii, viii, ix, 7, 35, 36, 37, 39, 41, 43, 44, 46, 49, 85, 86, 87, 88, 89, 90, 91, 93, 96, 109, 110, 111, 114, 115, 116, 119, 121, 124, 125, 127, 128, 129
contaminated sites, vii, 37, 50, 54, 55, 62
contaminated soil, 54, 111
contaminated water, 86
contamination, viii, ix, 35, 36, 37, 43, 46, 49, 51, 54, 58, 63, 79, 80, 87, 88, 91, 92, 93, 94, 95, 97, 109, 110, 111
contour, 150
contour plots, vii, 1
convergence, 80, 81, 82, 89, 91
convergence criteria, 81
cooling, 85
copolymers, 170, 179
copper, 149
correlation, 72
correlations, viii, 13, 178
corrosion, 55
cosmetics, 145
cost, ix, x, 36, 47, 79, 80, 81, 82, 86, 87, 88, 89, 90, 91, 92, 94, 95, 141, 142, 143, 144, 146, 163, 170
cost saving, 92
crude oil, vii, x, 141, 142, 144, 147, 158, 160, 161, 163, 169, 170, 178, 179
crystalline, 85, 154

D

decay, 110, 125
decomposition, 140
decoupling, 149
deficiency, 43
deformation, 14
Delta, 129
Department of Energy, 167
deposition, viii, ix, 10, 65, 66, 68, 72, 73, 75, 99, 100, 102, 104, 105
depression, 54
depth, 2, 3, 43, 67, 92, 114, 115, 142
derivatives, 80, 81, 84, 85, 110, 178
detection, vii, 87, 88, 94, 97
deviation, 4, 113, 117
diesel fuel, 170
differential equations, ix, 131, 132, 134, 136, 139, 140
diffraction, 146, 154
diffusion, 4, 5, 11, 38, 113
diffusion time, 5
diffusivity, 5, 22, 129
dilution of produced water, vii, 1
dimensionality, 81

discharge scenarios, vii, 1, 6, 10
discharges, 2, 5, 6, 8, 11
discontinuity, 82
discrete variable, 81
discretization, 110
dispersion, vii, ix, 1, 4, 5, 10, 38, 86, 109, 110, 111, 112, 113, 116, 117, 122, 124, 127, 128
displacement, 170
distilled water, 146, 156, 158, 171
distribution, viii, ix, 2, 5, 6, 7, 8, 36, 37, 39, 53, 54, 55, 60, 61, 62, 89, 110, 114, 116, 117, 118, 119, 120, 121, 122, 123, 124, 125, 126, 127, 158, 177
distribution function, 37, 39
divergence, 38
dosage, 126, 173
drainage, 14, 159, 170
drinking water, ix, 54, 109, 110, 111, 125, 126, 128
dynamic viscosity, 22, 59, 60

E

ecological risk, vii, 1, 2, 3, 4, 6, 7, 8, 9, 10, 11
editors, 127
effluent, 4, 5, 6, 178
Egypt, 142
election, 85
electric field, 148, 170, 179
electroflotation, 178
electron, 146, 147
electron microscopy, 146, 147
e-mail, 1, 99
emission, 149
emulsions, 169, 170, 171, 172, 177, 178, 179
enamel, 162
energy, 14, 22, 23, 64, 85, 146, 148, 149, 154, 167
enforcement, 36
engineering, vii, viii, 14, 83, 131, 140
Enhanced Oil Recovery, 142, 165, 166
entrapment, 144
entropy, 64
environment, ix, x, 2, 5, 43, 44, 50, 109, 127, 141, 142, 145, 146, 161, 163
environmental conditions, 42, 46, 47, 143
environmental effects, 170
environmental impact, 10, 143, 146
environmental management, 90
Environmental Protection Agency (EPA), 10, 51, 129
environmental quality, 36
environmental standards, 36
equilibrium, 55, 57, 62, 76, 90, 107, 111
equipment, 88
error estimation, 39

estimation process, 58
ethanol, 128
Europe, 76, 107, 142
evapotranspiration, 88
evolution, 80, 83, 84, 90, 111
examinations, 43
excitation, 149
execution, 39
experimental condition, 148
exploitation, 169
exposure, ix, 3, 5, 6, 7, 37, 50, 91, 110, 111, 115, 116, 119, 120, 121, 122, 123, 124, 125, 126, 127, 128
extraction, viii, x, 43, 53, 58, 59, 61, 63, 86, 89, 93, 94, 142, 146, 151, 152, 155, 158, 161, 163

F

fabrication, 170
field trials, 144
films, 144, 170
filtration, 22, 156, 170, 178
financial, 75
financial support, 75
fish, 3, 7, 8, 9
fitness, 91
fixed costs, 82, 88, 92
flexibility, 39, 88
flocculation, 147, 170, 173, 178
flooding, viii, x, 65, 66, 70, 75, 76, 100, 141, 142, 143, 144, 146, 156, 157, 158, 163
flotation, 170, 178
fluctuations, 54, 55
fluid, vii, 13, 14, 15, 16, 18, 19, 22, 23, 24, 25, 26, 29, 32, 37, 43, 55, 56, 58, 59, 60, 63, 67, 86, 100, 102, 120, 140
fluidized bed, 156
force, 85, 150, 151, 159
formaldehyde, 170
formation, vii, viii, ix, x, 2, 14, 16, 17, 18, 19, 20, 21, 23, 26, 32, 54, 65, 66, 68, 70, 72, 74, 75, 76, 99, 100, 101, 102, 104, 106, 141, 157, 171, 172, 177
formula, 102
fractal dimension, 139
frequency distribution, 121, 126
freshwater, 128
friction, 159
function values, 84
funding, 32, 127, 163
fuzzy membership, 36
fuzzy relation analysis, viii, 35, 36, 37, 39, 46, 50
fuzzy set theory, 36
fuzzy sets, 36

G

Genetic Algorithms optimisation techniques, ix, 79
genetics, 84
geometry, 4, 148
Germany, 165
glasses, 139
global demand, 145
glycine, 149
glycol, 173, 177
grain size, 89
graph, 29, 161
gravity, 14, 38, 54, 158
grids, 43
groundwater, vii, viii, ix, 35, 36, 39, 42, 43, 47, 49, 50, 51, 53, 54, 55, 58, 59, 62, 63, 79, 80, 81, 82, 83, 84, 85, 86, 87, 88, 89, 90, 91, 92, 93, 94, 95, 96, 97, 98, 110, 111, 113, 115, 117, 119, 122, 124, 128, 129
growth, 3, 7, 152, 159, 161, 163
guidelines, 36, 37, 43, 125, 128

H

half-life, 125
Hazard Quotients (HQs), vii, 1
hazardous waste, ix, 50, 51, 109, 110
hazards, 124
health, ix, 36, 45, 46, 49, 51, 54, 109, 110, 111, 115, 119, 121, 122, 123, 124, 125, 126, 127, 142, 145
health care, 142
health effects, ix, 109, 111, 126, 127
health problems, 123, 127
health risks, ix, 36, 109
heat transfer, 132, 139, 140
heavy metals, 10
height, 3
heterogeneity, 36, 143
high pH alkaline solution, x, 141
history, 19, 31, 32, 42, 58
human, ix, 36, 37, 47, 54, 109, 110, 111, 112, 125, 126, 127, 128, 142, 145, 162
human exposure, 111, 128
human health, 47, 54, 125, 126, 142, 145
Hunter, 95, 96, 166
hybrid, 37, 178
hydrocarbons, vii, viii, 35, 51
hydrogen, 154
hydroxide, x, 141, 143, 145, 146, 154, 156, 157, 161, 163
hydroxyl, 163

I

identification, 36, 62, 95, 98
image, 147, 148, 150, 151, 152
images, 147, 148, 149
imbalances, 142
incidence, ix, 99, 101, 102
India, 167, 169
individuals, ix, 84, 109, 111, 121, 126
Indonesia, 1
industrial chemicals, ix, 109, 110
industrial revolution, 142
industries, x, 145, 150, 169
industry, vii, 144, 145, 146, 161, 163
ingestion, ix, 110, 111, 112, 115, 119, 121, 122, 123, 124, 125, 126, 127
ingredients, 162
inhibition, 3
initiation, 111
injuries, 47, 145
integration, 2, 56, 57, 135
interface, x, 54, 58, 141, 144, 149, 158, 170
interfacial tension (IFT), x, 141
inversion, 98, 149
ions, 100, 148, 154, 155
Iran, 131
iris, 129
Islam, v, vi, x, 13, 33, 79, 141, 142, 143, 144, 145, 146, 154, 155, 161, 164, 165, 166
isolation, 93
issues, 80, 83
iteration, 140, 148

J

Japan, 10, 148

K

kidneys, 126
kinetic model, viii, 65
kinetics, 76, 178
KOH, 155

L

lakes, 5
landfills, ix, 109, 110
Latin America, 107
laws, 5
lead, 39, 45, 100, 171

leakage, 36, 42, 49, 145
lifetime, 120
light, 37, 54, 64, 147
linear programming, 90, 93
liquids, 64, 150, 170
liver, 126
LNAPL residing, viii, 53
longevity, 155
Louisiana, 11
lubricating oil, 170

M

magnesium, x, 142, 152, 154, 161
magnet, 149
magnetic resonance, 146
magnitude, ix, 27, 36, 38, 54, 65, 67, 92, 100
majority, 14, 54
management, viii, ix, 10, 35, 36, 49, 51, 79, 80, 81, 82, 83, 85, 86, 87, 88, 89, 90, 91, 93, 94, 96, 97, 98, 110
manganese, x, 142, 154
manipulation, 56
marine environment, 1, 5
MAS, x, 142, 149, 154, 155
mass, 37, 57, 64, 67, 91, 97
material balance equation (MBE), vii, 13
materials, 2, 148, 151, 162
mathematical programming, 82, 88, 93, 95
matrix, 40, 42, 46, 57, 88, 89
measurement, 7, 36, 82, 86, 93, 150
measurements, viii, 4, 13, 149, 160
media, viii, 15, 22, 37, 38, 50, 51, 53, 54, 55, 56, 61, 62, 63, 64, 65, 67, 68, 89, 104, 105, 140
membership, 36, 40, 41, 42, 50
membranes, 145, 178
memory, vii, 13, 14, 22, 23, 33
memory function, 14
metals, 10
meter, 3, 146
methodology, ix, 49, 55, 91, 109, 110, 111, 124, 127
Mexico, 76
Miami, 63
microscope, 147, 149
microscopy, 146, 147
migration, 54, 67, 97, 128
milligrams, 115
MIP, 93
mixing, 2, 3, 10, 66, 100, 102, 113
modelling, viii, 35, 84, 85, 89, 93, 178
models, viii, ix, 5, 6, 25, 38, 39, 44, 60, 65, 66, 70, 71, 72, 74, 75, 81, 83, 85, 87, 89, 91, 92, 93, 96, 97, 99, 110, 111, 114, 128

modules, 3
molecular weight, 173
molecules, 85, 159, 177
momentum, 4
Monte Carlo simulation, vii, ix, 1, 5, 8, 10, 36, 37, 39, 109, 110, 112, 116, 119, 121
morphology, 151
motivation, 85
mucous membrane, 145
multiple factors, viii, 35
mutation, 80, 84, 91
mutations, 85

N

natural evolution, 84
natural gas, vii, 42, 58
natural selection, 80, 84
neglect, 21
Nepal, 165
Netherlands, 51, 98
neural network, 89, 95, 97, 98
New South Wales, 11
Nigeria, 65, 76, 99, 107
nodes, 57
non-linear equations, 132
normal distribution, 125
North America, ix, 109, 110, 125, 128
North Sea, 178
Nuclear Magnetic Resonance (NMR), x, 141, 142, 146, 149, 154, 155, 163, 164, 166
nucleation, 104
nucleus, 149
nutrient, x, 142, 161, 163
nutrients, 152, 161, 163

O

obstacles, ix, 109
oceans, 5
oil, vii, viii, x, 1, 2, 13, 14, 15, 16, 17, 18, 19, 20, 21, 23, 24, 28, 29, 32, 53, 54, 55, 56, 57, 58, 60, 61, 62, 63, 76, 101, 104, 141, 142, 143, 144, 147, 148, 149, 156, 157, 158, 159, 160, 161, 162, 163, 164, 169, 170, 171, 172, 173, 174, 175, 176, 177, 178, 179
oil estimation analysis, viii, 53, 58
oil production, 16, 23, 32, 143, 171
oil spill, 55, 63, 178
Oklahoma, 76, 77, 107
olive oil, 162
operating costs, 92

operations, 75, 170
optimal resource allocation, 91
optimization, vii, 80, 81, 82, 83, 84, 85, 86, 87, 88, 89, 90, 91, 92, 93, 94, 95, 96, 97, 98
optimization method, 80, 81, 82, 83, 84, 85, 90, 93, 98
organic compounds, 154
oxygen, x, 142, 152
ozone, 170, 179

P

palladium, 148
parallel, 90, 98, 150
parallelization, 39
parameter estimation, 88, 89, 94, 96
partition, 59, 114, 125, 126, 127
pathways, 115, 142
peer review, 49
percentile, 7
performance measurement, 82
permeability, vii, ix, 13, 38, 45, 50, 65, 66, 69, 70, 71, 72, 73, 74, 75, 76, 99, 100, 101, 103, 104, 106, 110, 129, 157
pesticide, 50
petroleum, v, vii, viii, ix, x, 2, 10, 14, 32, 33, 35, 36, 37, 42, 49, 50, 51, 53, 54, 55, 62, 63, 65, 75, 76, 79, 80, 107, 108, 109, 110, 124, 127, 129, 144, 145, 147, 161, 164, 165, 166, 169, 170, 178
petroleum contaminants, ix, 79
phosphorus, 161
physical properties, 143, 158
plant growth, 161, 163
plants, x, 11, 142, 145, 146, 152, 161, 163, 170
platform, 5, 11
polarization, 149
policy, 85, 89, 90, 94
pollutants, 39, 48, 49, 96
pollution, 50, 51, 80, 87, 97, 110, 128, 145, 170
polymer, x, 141, 178
polymer solutions, x, 141
polypropylene, 170
population, ix, 80, 84, 85, 92, 109, 112, 124, 125, 127
population size, 84
porosity, vii, 13, 16, 38, 45, 56, 57, 59, 66, 68, 69, 70, 71, 72, 100, 101, 103, 105, 106, 113, 114, 117
porous media, viii, 22, 37, 38, 50, 51, 53, 54, 55, 56, 61, 62, 63, 64, 65, 66, 67, 68, 89, 104, 105, 140
potassium, x, 142, 146, 152, 154, 161
power plants, 146
precipitation, viii, ix, 65, 66, 72, 76, 99, 100, 101, 102, 157
preparation, 58, 171
pressure gradient, 68, 74, 75, 101
prevention, 66, 162
probability, vii, 1, 5, 6, 36, 37, 39, 55, 85, 87, 88, 93, 117, 118, 120, 121, 126, 128
probability distribution, vii, 1, 5, 6, 36, 37, 39, 118, 120, 126, 128
probability theory, 36
probe, 149
producers, 100
professionals, 110
programming, 81, 82, 86, 87, 88, 90, 91, 92, 93, 95, 96, 97, 98
project, 32, 66, 163
propagation, 50
public health, 51, 110
pulp, 145

Q

quantification, 36, 49
questionnaire, 43

R

radial distance, 67, 70, 72, 73, 100, 104, 105, 106
radius, 8, 9, 66
rainfall, 85
random numbers, ix, 110, 127
reactions, 144, 155
reactivity, 144
real numbers, 84
reality, 27
reasoning, 51
reciprocity, 96
recombination, 91, 92
reconstruction, 111
recovery, vii, viii, x, 13, 14, 17, 23, 29, 31, 32, 53, 54, 55, 58, 59, 60, 61, 62, 63, 89, 97, 141, 142, 143, 144, 147, 149, 156, 157, 164, 170
recovery factor (RF), viii, 13
recovery process, viii, x, 53, 63, 141, 142, 144
redundancy, 95
registry, 111, 164
regression, 5, 89
regression analysis, 89
regulations, 126
relative toxicity, 111
relaxation, 85
reliability, 87, 88, 91, 94, 127

remediation, viii, ix, 36, 43, 47, 48, 49, 50, 51, 53, 54, 55, 61, 62, 63, 80, 82, 85, 86, 87, 88, 89, 90, 91, 92, 93, 94, 95, 96, 97, 98, 110
renaissance, 142
requirements, 58, 80, 87
researchers, 14, 24, 81, 82, 83, 85, 93, 110, 144, 148, 155
residuals, 54
resins, 170
resolution, 155
resource allocation, 91
resource management, 10
resources, 80, 83, 85, 90, 92
response, 6, 88, 89, 177
restoration, 54
restrictions, 80
retardation, 113, 114, 116, 125
risk, vii, viii, ix, 1, 2, 3, 4, 6, 7, 8, 9, 10, 11, 35, 36, 37, 39, 40, 41, 42, 43, 44, 45, 46, 47, 48, 49, 50, 51, 90, 91, 94, 110, 111, 112, 115, 116, 119, 121, 122, 123, 124, 125, 126, 127, 145
risk assessment, viii, 2, 10, 11, 35, 36, 37, 39, 41, 44, 46, 47, 48, 49, 50, 51, 91, 124, 127
risk management, 51, 90
risk profile, 10
room temperature, 171, 173
rules, 81, 83, 84
runoff, 85

S

safety, 47, 110, 128, 142
salinity, 2
salt concentration, 75, 101, 102
salts, 75, 144, 155
saturation, 15, 16, 18, 21, 27, 38, 51, 55, 56, 57, 58, 59, 66, 67, 70, 71, 100, 101, 103, 104, 106
savings, 92
scaling, 56, 60, 75, 107
scanning electron microscopy, 146
Schrödinger equation, 132, 139
science, 142
search space, 85
seed, 162
SEM micrographs, x, 141
sensitivity, 95, 150, 162
set theory, 36
sewage, 5
shape, x, 26, 28, 101, 102, 142
shear, 22
shellfish, 3
showing, 160, 163
shrimp, 3

silica, x, 142, 154
silicon, x, 142, 152
simulation, vii, viii, ix, 1, 5, 6, 10, 14, 25, 36, 37, 39, 43, 50, 53, 55, 60, 61, 62, 75, 83, 86, 87, 88, 89, 90, 92, 93, 95, 96, 98, 109, 110, 111, 116, 119, 121, 127
simulations, vii, 1, 5, 7, 8, 9, 11
skin, 66, 73, 74, 145
smoking, 162
smoothing, 159
sodium, x, 141, 142, 143, 145, 146, 152, 156, 157, 161, 163
sodium hydroxide, x, 141, 143, 145, 146, 156, 157, 161, 163
software, vii, 1, 7, 10, 70, 75, 90, 147, 150
soil type, 43
solid state, 67
solubility, 18, 45, 54, 59, 116, 125, 154
solution, ix, x, 15, 18, 64, 80, 81, 82, 83, 84, 85, 87, 89, 91, 93, 97, 109, 110, 114, 127, 131, 134, 140, 141, 142, 143, 144, 145, 149, 151, 152, 154, 156, 157, 158, 159, 160, 161, 162, 163
solution space, 82
sorption, 90, 96
Spain, 76, 107
speciation, 154
species, 3, 6, 154, 166
spectroscopy, 149
spin, 149
stability, x, 110, 169, 170, 171, 172, 173, 177
stabilizers, 170
standard deviation, 4, 113
state, vii, ix, 13, 50, 67, 79, 81, 83, 85, 92, 93, 101, 110, 113, 114, 124, 164, 178
states, 126, 149, 170
statistic test, 161
statistics, 95, 127
stochastic model, 50
stochastic processes, 39
storage, viii, ix, 41, 42, 50, 53, 54, 58, 63, 80, 109, 110, 111, 124, 127, 129
stress, vii, 13, 14, 22, 23, 81
structure, 39, 98, 148
sulfate, 66
Sun, 139
Superfund, 87, 129
suppression, 150
surface layer, 3
surface structure, 148
surface tension, 22, 59, 166
surfactant, 50, 63, 158, 159, 164
surfactants, x, 141, 170
survival, 3, 7, 8, 9, 84

T

suspensions, 170
swelling, 144
synergistic effect, 170, 179

Taiwan, 10
tanks, viii, ix, 41, 50, 53, 54, 58, 63, 109, 110, 129
target, 142, 149
techniques, vii, ix, x, 10, 13, 14, 37, 39, 49, 51, 79, 80, 81, 82, 83, 84, 85, 87, 88, 89, 90, 92, 93, 94, 96, 139, 141, 142, 146, 147
technology, x, 54, 55, 63, 169
teeth, 162
temperature, vii, 2, 13, 22, 67, 68, 70, 71, 85, 100, 103, 104, 106, 170, 171, 173, 175, 177, 178
tension, x, 22, 45, 59, 141, 142, 144, 150, 151, 160, 161, 163, 166, 172, 173
testing, 111, 146
texture, 116, 151
thermal expansion, 22
thermodynamic method, 139
thermodynamics, viii, 65
threats, 44
time dependent porosity, vii, 13
time series, 5
tissue, 145
titanium, x, 142, 152
toluene, ix, 43, 45, 109, 111, 112, 116, 117, 118, 120, 121, 122, 123, 124, 125, 126, 127
tooth, 162
toxicity, 6, 7, 111
trade, 87, 88
trade-off, 87, 88
transmission, 42, 58
transport, viii, ix, 35, 37, 39, 43, 44, 50, 58, 81, 83, 86, 87, 88, 90, 91, 93, 96, 98, 109, 110, 111, 112, 124, 127, 128, 129
transportation, 142
treatment, x, 88, 89, 96, 145, 152, 161, 169, 171, 178
trial, 84, 91
Trinidad, 76, 107
Trinidad and Tobago, 107
turbulence, x, 4, 169, 177

U

uniform, 15, 39, 54, 63, 114
unique features, 32
united, 107, 108, 110, 128, 129, 146, 161, 165
United Kingdom (UK), 107, 108, 148
United Nations, 128

United States, 110, 129, 146, 161, 165
universal gas constant, 22
urine, 111
USA, 76, 107, 143, 144, 146, 164, 165

V

vacuum, viii, 43, 53, 54, 55, 59, 60, 61, 62, 148
vacuum-enhanced multiphase extraction, viii, 53, 61, 63
validation, 70
vapor, 54, 60
variable costs, 91
variables, ix, 39, 50, 51, 79, 81, 82, 83, 88, 89, 90, 92, 93, 97, 101, 109, 112, 119, 121, 123, 124, 125, 127
variations, ix, 3, 28, 45, 109, 110, 112, 127
vector, 38, 150
vegetable oil, 162
vehicles, 142
velocity, 3, 22, 38, 67, 70, 100, 104, 110, 113, 114, 116, 117
VER modeling results, viii, 53, 63
viscosity, 22, 38, 59, 60, 67, 70, 100, 104, 106, 171, 177
volumetric changes, 15

W

Wales, 11
Washington, 11, 51, 129
waste, viii, ix, 2, 35, 36, 42, 49, 50, 51, 58, 88, 93, 109, 110, 111, 170
waste management, viii, 35, 36, 49
wastewater, x, 169, 170, 171, 178, 179
water quality, 10, 90, 98, 128, 170
water resources, 80, 83, 85, 90
water supplies, 110, 111
weakness, 2, 127
wells, 25, 55, 58, 60, 63, 82, 86, 87, 88, 90, 92, 100, 101, 104, 113
West Indies, 76, 107
wettability, 144
wetting, 114, 115, 117
White Rose field, vii, 1, 10
withdrawal, 29, 30, 31
wood, x, 141, 142, 144, 146, 148, 149, 151, 152, 153, 154, 155, 156, 157, 158, 159, 160, 161, 162, 163
wood ash, x, 141, 142, 144, 146, 148, 149, 151, 152, 153, 154, 155, 156, 157, 158, 159, 160, 161, 162, 163

World Health Organization (WHO), 128
worldwide, 142, 145, 146, 161

X

X-ray analysis, x, 142, 147
XRD, x, 141, 142, 146, 149, 154, 163, 166

Y

yield, 88, 98

Z

Zimbabwe, 164